Urban Renewal, Governance and Sustainable Development: More of the Same or New Paths?

Urban Renewal, Governance and Sustainable Development: More of the Same or New Paths?

Editor

Ingemar Elander

MDPI • Basel • Beijing • Wuhan • Barcelona • Belgrade • Manchester • Tokyo • Cluj • Tianjin

Editor
Ingemar Elander
Örebro University
Sweden

Editorial Office
MDPI
St. Alban-Anlage 66
4052 Basel, Switzerland

This is a reprint of articles from the Special Issue published online in the open access journal *Sustainability* (ISSN 2071-1050) (available at: https://www.mdpi.com/journal/sustainability/special_issues/Urban_Renewal).

For citation purposes, cite each article independently as indicated on the article page online and as indicated below:

LastName, A.A.; LastName, B.B.; LastName, C.C. Article Title. *Journal Name* **Year**, *Volume Number*, Page Range.

ISBN 978-3-0365-4271-3 (Hbk)
ISBN 978-3-0365-4272-0 (PDF)

© 2022 by the authors. Articles in this book are Open Access and distributed under the Creative Commons Attribution (CC BY) license, which allows users to download, copy and build upon published articles, as long as the author and publisher are properly credited, which ensures maximum dissemination and a wider impact of our publications.
The book as a whole is distributed by MDPI under the terms and conditions of the Creative Commons license CC BY-NC-ND.

Contents

About the Editor vii

Ingemar Elander
Urban Renewal, Governance and Sustainable Development: More of the Same or New Paths?
Reprinted from: *Sustainability* **2022**, *14*, 1528, doi:10.3390/su14031528 1

Kristian Hoelscher, Hanne Cecilie Geirbo, Lisbet Harboe and Sobah Abbas Petersen
What Can We Learn from Urban Crisis?
Reprinted from: *Sustainability* **2022**, *14*, 898, doi:10.3390/su14020898 9

Mikael Granberg and Leigh Glover
The Climate Just City
Reprinted from: *Sustainability* **2021**, *13*, 1201, doi:10.3390/su13031201 25

Llewellyn Leonard and Rolf Lidskog
Conditions and Constrains for Reflexive Governance of Industrial Risks: The Case of the South Durban Industrial Basin, South Africa
Reprinted from: *Sustainability* **2021**, *13*, 5679, doi:10.3390/su13105679 45

Jörgen Johansson and Jonas Gabrielsson
Public Policy for Social Innovations and Social Enterprise—What's the Problem Represented to Be?
Reprinted from: *Sustainability* **2021**, *13*, 7972, doi:10.3390/su13147972 65

Terence Fell and Johanna Mattsson
The Role of Public-Private Partnerships in Housing as a Potential Contributor to Sustainable Cities and Communities: A Systematic Review
Reprinted from: *Sustainability* **2021**, *13*, 7783, doi:10.3390/su13147783 85

Bozena Guziana
Only for Citizens? Local Political Engagement in Sweden and Inclusiveness of Terms
Reprinted from: *Sustainability* **2021**, *13*, 7839, doi:10.3390/su13147839 111

Marco Eimermann, Urban Lindgren and Linda Lundmark
Nuancing Holistic Simplicity in Sweden: A Statistical Exploration of Consumption, Age and Gender
Reprinted from: *Sustainability* **2021**, *13*, 8340, doi:10.3390/su13158340 137

Jasmina Nedevska
An Attack on the Separation of Powers? Strategic Climate Litigation in the Eyes of U.S. Judges
Reprinted from: *Sustainability* **2021**, *13*, 8335, doi:10.3390/su13158335 153

About the Editor

Ingemar Elander

Ingemar Elander is professor in politics at Örebro University and is associated with Mälardalen University, Sweden. His research interests cover urban governance in a broad sense, as exemplified in several publications on cities and climate change, environment and democracy, faith-based organisations and politics, urban partnerships, and public health. Recent research covers (un)sustainable development and urban renewal in Swedish neighbourhoods and cities, with a special focus on social and political dimensions as documented in publications co-authored with geographer Eva Gustavsson (2009; 2016, 2019). He is a co-editor of Urban Governance in Europe (Eckardt & Elander 2009), and co-author of Faith-based Organisations and Social Exclusion in Sweden (Elander & Fridolfsson 2011). The latter is an extensive report which has inspired several reviewed articles and book chapters authored with Charlotte Fridolfsson 2012–2021. Recently published articles and work in progress include co-authored manuscripts on securitization aspects related to migration, COVID-19, climate change, and the development of central-local government relations and planning in Sweden.

Editorial

Urban Renewal, Governance and Sustainable Development: More of the Same or New Paths?

Ingemar Elander

School of Humanities, Education and Social Sciences, Örebro University, 70281 Örebro, Sweden; ingemar.elander@oru.se

Citation: Elander, I. Urban Renewal, Governance and Sustainable Development: More of the Same or New Paths? *Sustainability* **2022**, *14*, 1528. https://doi.org/10.3390/su14031528

Received: 17 January 2022
Accepted: 20 January 2022
Published: 28 January 2022

Publisher's Note: MDPI stays neutral with regard to jurisdictional claims in published maps and institutional affiliations.

Copyright: © 2022 by the author. Licensee MDPI, Basel, Switzerland. This article is an open access article distributed under the terms and conditions of the Creative Commons Attribution (CC BY) license (https://creativecommons.org/licenses/by/4.0/).

Humanity seems to have been thrown into a 'perfect storm' of several huge challenges such as global warming, accelerating extinction of species, the corona pandemic and uncontrollable migration streams caused by fossil fuel emissions, overexploitation of natural resources, extreme weather, viruses, and ethnic and religious conflicts. On top of this, there are even signs that liberal democracy is in crisis, far from the days when it was proclaimed irreversibly hegemonic [1]. The challenges mentioned are, by many scientists, some world leaders and a broader audience, considered existential threats in need of urgent action, that is 'securitization' [2]. As the causes, effects and adequate reactions upon these threats are contested, there are no given solutions how to 'de-securitize' them, neither one by one, no less together. In other words, it is ultimately a question of how government and governance configurations globally, at various levels and settings, choose to decide on the road forward. Although world leaders in the Global North have been desperately fertilizing their economies with heaps of money to recover the still largely fossil-dependent 'Great Acceleration', there are in this 'critical juncture' also windows of opportunity to enter new paths. How do urban public institutions and actors in market and civil society respond to these crises? Do they only try to reinvent old ideas and practices, or do they search for healthier, more sustainable, democratic and just ways of handling major threats and risks?

There is a widening gap between current and expected development and what needs to be done in responding to climate change: 'Record atmospheric greenhouse gas concentrations and associated accumulated heat have propelled the planet into uncharted territory, with far-reaching repercussions for current and future generations' [3], The Climate Pact agreed upon at Glasgow in October 2021 ended up with representatives of the participating nations signing a deal calling for the world's major emitters of greenhouse gases to keep the key target of limiting the increase of global warming to 1.5 degrees as agreed upon in Paris 2015. However, as the increase in temperature since then has already reached 1.2 degrees, this seems unlikely, as it implies that most of the remaining oil, gas and coal reserves on earth have to remain unexploited [4]. At least the Pact promises more money for developing nations to help them adapt to the worst effects of climate change, also clarifying the rules around carbon markets. At the very end, however, India, backed by China and other coal-dependent nations, in the Global South rejected a clause calling for the 'phase out' of fossil-fired power in favor of the less demanding wording 'phase down'. The implication of that switch of one single word was met dismally by many EU countries, and in particular all small island nations, who are the most vulnerable to rising sea levels caused by climate change. As succinctly stated by the Guardian columnist George Monbiot: World leaders 'make a series of grand announcements, then convince us they have saved the planet. Words are politically cheap, actions are expensive' [5].

The multilevel handling of climate-change mitigation and adaptation reflects the complexity of this policy area. Nation-states and other actors such as trans-national companies, universities, scientific laboratories, NGOs and social movements are all important members in the choir of voices needed to phase out fossil-fired power dependency. Agreements and regulations at global, international and national levels are, indeed, important frameworks within which regional and local policy measures are taken, but action by public, private, and voluntary association actors may also influence policy making at, or between, various levels. Whether these actions will also have an impact on high-level, large-scale decisions is questionable, although they are certainly crucial when it comes to local climate change adaptation in relation to the *consequences* of extreme weather. The sheer magnitude of policy initiatives at the local level, their diversity, and the experimental and practical nature of many local projects, are bound to bring forward genuinely new ideas and solutions, but to have an overall impact, they also have to be a matter of national, international and global politics, not to forget the potential of a massive bottom-up mobilization by youth and adult climate activists in civil society.

Although we have a multi-level governance system where issues of responsibility and respect for future generations and non-human species are discussed and, at least to some extent, taken into account, this is far from a rosy picture of different actors getting together in search for efficient and democratic measures to mitigate and adapt to climate change. On the contrary, there is an ongoing *struggle* among actors on different levels concerning the right to take part in defining and addressing global issues, and there is no guarantee that policies will move in the direction of a broad, final consensus on the road towards a fossil-free planet. There is even a world-wide strong coal-lobby as brought into the light by the 'climategate' [6], including powerful politicians like West Virginia Democratic Senator Joe Manchin, so far making Congress block the Biden administration 'Build Back Better' Act, including its long-term phasing-out-fossil-power agenda [7].

However—and not to forget—almost thirty years ago, in the wake of the 1992 Earth Summit in Rio, and the 1996 Habitat II Conference in Istanbul, the world experienced a worldwide, locally based movement addressing 'sustainable development'. The World Bank launched the Urban Partnership initiative, urging city and national officials to mobilize 'the resources and talents of bilateral organizations, NGOs, academics, corporations, foundations, and individuals to activate carefully selected teams of experts who will work with them to develop strategic frameworks and to chart pathways for long term growth'. The premise was that 'analysis of a successful city should not be one-dimensional, but must include all the elements of livability, productivity, competitiveness and governance': [8] 'The global threats—mainly resource depletion, environmental degradation and climate change—require global partnership. This global partnership has become a central issue for world peace. Global environmental and development policy is the peace policy of the future' [9].

Partnerships in the Habitat II context were then perceived to have 'a very wide scope, encompassing international capacity building programs at the national and subnational levels'. The aim was to empower all interested parties, particularly local authorities, the private sector, the co-operative sector, trade unions, non-governmental and community-based organizations, to enable them to play an effective role in shelter and human settlements planning and management '/ ... /Each Government should ensure the right of all members of its society to take an active part in the affairs of the community in which they live and ensure and encourage participation in policymaking at all levels' [10].

Common arguments in favor of partnership mentioned in the policy-orientated literature then revolved around the concepts synergy, transformation, and budget enlargement, often exemplified by referring to the joint venture between a profit-seeking commercial firm and non-profit organizations like a local government or a charity. An associated negotiation process over the distribution of profits was then perceived to favor private shareholders as well as social ends. However, even at the time of launching the partnership, vision critical voices reminded us that 'partnerships are complex systems with considerable potential but at the same time full of unexpected traps for the unwary' [8].

Gradually, concepts like 'ecological citizen', 'political consumer' and 'moral agent' became commonplace, thus signaling the potential and willingness of individuals and households to change their attitudes and behavior towards sustainable consumption and lifestyles [11,12]. In addition, Extinction Rebellion, Fridays for Future and other action initiatives are examples of a world-wide bottom-up movement with growing potential. However, green values, attitudes, and a willingness to change one's consumption behavior do not just emerge out of talk and protest. Inhabitants of the earth live in a global system of capitalism based on fossil-fuel dependent production and consumption that leaves no one free to choose an alternative, but in small pockets of knowledge, time and space. In particular, most people living in the Global South are, during the foreseeable future, strongly dependent on fossil-fuel-based production of energy to increase their living standard, regardless of global agreements on net-zero emissions of CO_2. In other words, there is a long way to go from pockets of 'green growth' to potentially 'de-growth' societies [13,14]. Still, long-term change in behavior is possible, although in need of substantial support from responsible multi-level, public and private institutions. Considering the complexity of global warming and sustainable development in all their aspects, there is not, and could not be, one ultimate governance fix for securitizing the consequences of climate change related threats. What we have is a patchwork of partly overlapping assemblies, located at different levels and sectors, and representing different spheres of authority. Government institutions establish links to the parallel structures of informal, voluntary associations such as social movements and environmental associations, as well as individuals, households and for-profit companies, that is a system of 'hybrid governance' [2,15].

Although participation and deliberation within the framework of representative institutions could be supportive in the struggle for a low carbon dependent, sustainable society, these mechanisms are still being used for practices dependent on fossil fuel-based economic growth and an ever-increasing excessive consumption in the Global North. Despite this, there is no lack of local counter-initiatives in favor of creating low carbon, sustainable cities and neighborhoods, signposted by catchwords such as 'eco-cities', 'low-carbon cities', 'sustainable cities', 'zero carbon cities', 'green cities', 'just cities', 'transition towns', 'virtual cities', 'Oekostadt', 'healthy cities', 'inclusive cities', etc. In addition, 'slim', 'slow', 'de-growth', and 'Lo-Tek' cities indicate alternatives to the still strong techno-digital strand of urban renewal visions [11].

In its latest report, the Intergovernmental Panel on Climate Change provides an overview of the current state of knowledge and highlights the need for society to handle the huge dual and interlinked challenge of radically reducing the dependency on fossil fuels and preparing for future outbreaks of extreme weather [16]. *Mitigation* of climate change aims at halting the increase of carbon dioxide in the atmosphere and making a fossil-free future possible, whereas adaptation aims at making society more resilient to the impacts of climate change whatever its root causes, that is reducing damage caused by sudden threats like flooding, heatwaves, droughts and forest fires. The state of knowledge about the historical development and the current situation, as well as credible forecasts about what will happen, are scientifically well founded as a basis for policies and planned activities to mitigate the emissions causing climate change and adapt to contemporary and future impacts and risks, in the direction of making society more secure [17,18].

Unfortunately, however, there is no corresponding knowledge of exactly where, when and with what specific consequences global warming will appear. Nevertheless, mitigation and adaptation are both crucial aspects implying efforts at all levels and sectors of society, that is multi-level, hybrid governance and planning targeted at the reduction of greenhouse gas emissions and risk reduction [2]. The scale of these challenges calls for the state to be strongly involved in multi-scalar securitizing efforts that are coordinated with businesses, non-governmental organizations, citizens and inhabitants at national, regional and local levels. Thus, both mitigation and adaptation are collective security issues characterized by considerable uncertainty concerning policy and measures regarding who should take responsibility for those actions. Notably, with regard to adaptation to extreme events such as flooding, wild-fires and contaminated air, there is no escape, not even for those who still negate or ignore the real human-induced causes of global warming.

Climate change then has to be addressed in a policy framework taking a broad view, acknowledging economic, cultural, social and political dimensions and their complicated interrelationships with overall ecological concerns [19]. As once stated by Peter Marcuse: 'Sustainability is not enough/ . . . /the promotion of "sustainability" may simply encourage the sustaining of the unjust status quo and how the attempt to suggest that everyone has common interests in "sustainable urban development" masks very real conflicts of interest' [20]. Presuming an institutional (dis)order with crisscrossing spheres of governance configurations adapted to context-bound circumstances, it is no wonder that governance research has become strikingly multifaceted in terms of theory, method, and empirical focus. The eight contributions in this Special Issue exemplify innovative ideas and practices related to the challenges of 'sustainability', including climate change and its repercussions on urban settings.

The contribution by Hoelscher, Geirbo, Harboe and Petersen is a critical examination of narratives and practices of resilience and sustainability as applied in various urban settings, especially in the Global South. Arguing that common 'best practice' governance approaches may rather contribute to further crisis, they instead bring attention to 'reflexivity'—a capacity to reconsider core values and practices as crucial for learning by experience in urban policy and practice. Presuming the potential of crisis will become the 'new normal', they argue that policy and analysis of places should instead be examined and practiced with special attention to their implications for values such as 'justice' and 'equity'. Offering a conceptual representation of urban crisis in terms of capacity, necessity, and possibility as a three-dimensional lens, they finally exemplify their approach through snapshots from case studies of cities of Detroit (MI, USA) and Medellin (Colombia).

Granberg and Glover, in their study of three very different cities (Baltimore (MA, USA), Karlstad (Sweden) and Port Vila (Venezuela), highlight the need for addressing the socially unjust, power asymmetrical causes and consequences of how climate risks and vulnerabilities are distributed in terms of justice and power. What priorities and values are at stake? Who decides and how? Who pays and who gains? Although climate change adaptation has primarily been informed by meteorology and related disciplines, this discourse has widened also to include social sciences, subjecting adaptation practices to analysis in terms of policy, governance and implementation. In this article, adaptation of climate change is critically analyzed and discussed in the context of the 'just city' as a key theoretical and normative concept. The authors conclude that the social context with its power and justice asymmetries is crucial to understand the distribution and handling of climate risks and vulnerabilities.

In their study of the South Durban Industrial Basins, Lidskog and Leonard take 'reflexive governance' as their conceptual point of departure, stating that this is a 'new mode of governance', being more inclusive and efficient in responding to complex risks than other approaches. Arguing that there is limited scholarly work on the theoretical and empirical foundations of this governance approach, especially how it may unfold in the Global South, they explore the conditions and constraints for reflexive governance in one of the most polluted regions in southern Africa, also being the scene for contention between residents, industry and government, and the struggle for inclusive democracy during apartheid. The authors find constraints involved in enabling reflexive governance, the most important one being the exclusionary, close alliance between government and industry, overshadowing the protection of social and environmental values, thus again highlighting the need for conceptually informed, context-sensitive analysis.

Offering a multitude of interpretations, 'social sustainability' is a rather vague and misused concept, and therefore has to be meticulously specified to be more than just words in policy and planning practice. Drawing upon Carol Lee Bacchi's methodological approach, Johansson and Gabrielsson present results in their article from a context-sensitive study of proclaimed public policy for social innovations and enterprise in a Swedish city. Policy and research on perceived sustainability policies are loaded with ideological contradictions, here illustrated by projects undergoing policy shifts from stronger, collective sustainability ambitions to 'softer', individual-focused problem representations driven by pragmatic interpretations and the perceived need to make decisions based on a limited range of information. The article concludes by emphasizing the need for more critical reflection on how the social dimension is defined when going from words to deeds in implementing ideologically colored concepts like 'sustainable development´.

The article by Fell and Mattson follows up on a critical stance to the longstanding public–private partnership (PPP) hype in much literature on the phenomenon. Using the Doughnut Economics (DE) model as their analytical guide, they conduct a systematic literature review illuminating the shortcomings and limitations of PPP and its potential ability to grapple with unsustainable urban development. The results show that PPPs are often far from inclusive, and instead rather excluding local actors—'the people'—from collaborative participation. Still the model could be useful as a tool to test the scope and depth of local collaborations, for example in the context of international treaties like the Social Development Goals, however, preferably adding a fourth P—people—to the triple formula. They thus conclude that there is still a long way to go to make PPPs inclusive in terms of various dimensions of social justice.

The article by Guziana brings to the fore another gap in the context of participation policy, practice and research—often advertised under a 'social sustainability' flag—namely who are/are not accepted as citizens in a basic civic and political sense. Worldwide migration makes national policymaking stuttering by not considering the political and ethical concerns about immigrants being defined as citizens, non-citizens, 'nomads' or 'the others'. Taking Sweden as an example, the article demonstrates the lack of inclusive language at local government websites, thereby blurring online information on who are/are not invited to participate. Analysis of information on municipal websites thus reveals that the term *citizen* is commonly used in a way literally excluding all residents not having acquired formal citizenship. One normative conclusion then is that government authorities should evade using the term 'citizen' when addressing their public, except for instances explicitly requiring formal citizenship.

Taking their point of departure in the notion of 'holistic simplicity', Eimermann, Lindgren and Lundmark explore the willingness of people in the Global North to revise their lifestyles, for example by cultivating their own vegetables, spending more time with relatives, neighbors and friends, as well as reading, walking and doing other everyday activities not dependent on extensive pressure on natural resources. Applying this concept to examine lifestyle patterns in different demographic groups, the article contributes to scholarly debates on the prospects of entering paths towards a 'de-growth' society. Drawing upon quantitative register data as a novel method in this field, they demonstrate how to measure the relative magnitude of holistic simplifiers in a population. Selecting smaller samples for longitudinal studies would make it possible to identify individuals willing to reduce their income, move away from more affluent, densely populated places, reduce their negative impact on the environment and turning to 'downshifting' lifestyles.

Jasmina Nedevska in her 'hypothesis' contribution brings attention to the increasing role of courts in climate litigation as a relatively new, understudied tool in climate governance not least in cases where local and global green movements raise protest against projects causing further exploitation of natural resources and increase of CO_2 emissions. The author presents a plan for studying judges' opinions and dissents in the United States exploring the views of U.S. judges on climate litigation conflicts in relation to the ideal of separation of powers: directly, as a political question doctrine, and indirectly, as a doctrine of legal standing. Considering the increasing number of conflicts between the stated goals of CO_2 reduction and fossil fuel-dependent industry, the role of courts, not only in the United States, will become even more frequent and thus in need of critical studies.

List of Contributions
- Hoelscher, K.; Geirbo, H.C.; Harboe, L.; Petersen, S.A. What can we learn from urban crisis? Innovation, flexibility and new urban practice.
- Granberg, M.; Glover, L. The Climate Just City.

- Lidskog, R.; Leonard, L. Conditions and Constraints for Reflexive Governance of Industrial Risks: The Case of the South Durban Industrial Basin, South Africa.
- Johansson, J.; Gabrielsson, J. Public Policy for Social Innovations and Social Enterprise—What's the Problem Represented to Be?
- Fell, T.; Mattsson, J. The Role of Public-Private Partnerships in Housing as a Potential Contributor to Sustainable Cities and Communities: A Systematic Review.
- Guziana, B. Only for Citizens? Local Political Engagement in Sweden and Inclusiveness of Terms.
- Eimermann, M.; Lindgren, U.; Lundmark, L. Nuancing Holistic Simplicity in Sweden: A Statistical Exploration of Consumption, Age and Gender.
- Nedevska, J. An Attack on the Separation of Powers? Strategic Climate Litigation in the Eyes of U.S. Judges.

Funding: This research received no external funding.

Informed Consent Statement: Not applicable.

Data Availability Statement: Not applicable.

Conflicts of Interest: The authors declare no conflict of interest.

References

1. Freedom House. Freedom in the World 2021. Democracy under Siege. 2021. Available online: https://freedomhouse.org/report/freedom-world/2021/democracy-under-siege (accessed on 20 November 2021).
2. Elander, I.; Granberg, M.; Montin, S. Governance and planning in a 'perfect storm': Securitizing climate change, migration and Covid-19 in Sweden. *Prog. Plan.* **2021**, 100634. [CrossRef]
3. World Meteorological Association. State of Climate in 2021: Extreme Events and Major Impacts. 2021. Available online: https://public.wmo.int/en/media/press-release/state-of-climate-2021-extreme-events-and-major-impacts (accessed on 10 December 2021).
4. Climate Action Tracker. 2021. Available online: https://climateactiontracker.org/press/Glasgows-one-degree-2030-credibility-gap-net-zeros-lip-service-to-climate-action/ (accessed on 10 December 2021).
5. Monbiot, G. If I Sound Angry, It's Because I am the Guardian. 4 November 2021. Available online: https://mail.google.com/mail/u/0/#inbox/FMfcgzGlkjbSMMJZBXrrvqnFLdmjsvjr (accessed on 14 November 2021).
6. Grundmann, R. Climategate and the scientific ethos. *Sci. Technol. Hum. Values* **2011**, *38*, 67–93. [CrossRef]
7. Goodell, J. Manchin's Coal Corruption is so much worse than you knew. *Rolling Stone*. 10 January 2022. Available online: https://www.rollingstone.com/politics/politics-features/joe-manchin-big-coal-west-virginia-1280922/ (accessed on 16 January 2022).
8. Elander, I. *Urban Partnerships*; UNESCO; Blackwell Publishing: Hoboken, NJ, USA, 2002; The citation is taken from a document presented at the ISSC-UNESCO Conference in Paris, November 1997. The document is not available on the internet.
9. Töpfer, K. Perspectives on the future: From a northern government. In *The Way Forward: Beyond Agenda 21*; Dodds, F., Ed.; Earthscan: London, UK, 1997; pp. 238–244.
10. UNCHS (United Nations Centre for Human Settlements). Habitat Agenda and Istanbul Declaration. In Proceedings of the Second United Nations Conference on Human Settlements, Istanbul, Turkey, 3–14 June 1996; United Nations Department of Public Information: New York, NY, USA, 1996.
11. Gustavsson, E.; Elander, I. Behaving Clean without Having to Think Green? Local Eco-Technological and Dialogue-Based, Low-Carbon Projects in Sweden. *Urban Technol.* **2017**, *24*, 93–116. [CrossRef]
12. Nielsen, K.S.; Nicholas, K.A.; Creutzig, F.; Dietz, T.; Stern, P.C. The role of high-socioeconomic-status people in locking in or rapidly reducing energy-driven greenhouse gas emissions. In *Nature Energy*; Springer: Berlin/Heidelberg, Germany, 2021. Available online: https://www.nature.com/articles/s41560-021-00900-y.pdf (accessed on 13 October 2021).
13. Parrique, T.; Barth, J.; Briens, F.; Kerschner, C.; Kraus-Polk, A.; Kuokkanen, A.; Spangenberg, J.H. *Decoupling Debunked: Evidence and Arguments against Green Growth as a Sole Strategy for Sustainability*; European Environmental Bureau: Brussels, Belgium, 2019. Available online: https://eeb.org/library/decoupling-debunked (accessed on 13 October 2021).
14. Kallis, G.; Demaria, F.; D'Alisa, G. Degrowth. In *International Encyclopedia of the Social & Behavioral Sciences*; Wright, J.D., Ed.; Elsevier: Oxford, UK, 2015; Volume 6, pp. 24–30.
15. Vakkuri, J.; Johanson, J.-E. *Hybrid Governance, Organisations and Society. Value Creation Perspectives*; Routledge: London, UK, 2020.
16. IPCC. The Production Gap Report. 2021. Available online: https://productiongap.org/wpcontent/uploads/2021/10/PGR2021_web_rev.pdf (accessed on 20 October 2021).
17. Low, N. *Being a Planner in Society. For People, Planet, Place*; Edward Elgar: Cheltenham, UK, 2021.
18. Dryzek, J.; Pickering, J. *The Politics of the Anthropocene*; Oxford University Press: Oxford, UK, 2019.

19. Lidskog, R.; Elander, I. Addressing climate change democratically: Multi-level governance, transnational networks and governmental structures. *Sustain. Dev.* **2010**, *18*, 32–41. [CrossRef]
20. Marcuse, P. Sustainability is not enough. *Environ. Urban.* **1998**, *10*, 103–112. [CrossRef]

Article

What Can We Learn from Urban Crisis?

Kristian Hoelscher [1,*], Hanne Cecilie Geirbo [2], Lisbet Harboe [3] and Sobah Abbas Petersen [4]

[1] Peace Research Institute Oslo (PRIO), 0186 Oslo, Norway
[2] Department of Computer Science, Faculty of Technology, Art and Design, Oslo Metropolitan University, 0130 Oslo, Norway; Hanne.Cecilie.Geirbo@oslomet.no
[3] Institute of Urbanism and Landscape, The Oslo School of Architecture and Design, 0175 Oslo, Norway; lisbet.harboe@aho.no
[4] SINTEF Digital, 7034 Trondheim, Norway; Sobah.Petersen@sintef.no
* Correspondence: hoelscher@prio.org

Abstract: The irreversible transition towards urban living entails complex challenges and vulnerabilities for citizens, civic authorities, and the management of global commons. Many cities remain beset by political, infrastructural, social, or economic fragility, with crisis arguably becoming an increasingly present condition of urban life. While acknowledging the intense vulnerabilities that cities can face, this article contends that innovative, flexible, and often ground-breaking policies, practices, and activities designed to manage and overcome fragility can emerge in cities beset by crisis. We argue that a deeper understanding of such practices and the knowledge emerging from contexts of urban crisis may offer important insights to support urban resilience and sustainable development. We outline a simple conceptual representation of the interrelationships between urban crisis and knowledge production, situate this in the context of literature on resilience, sustainability, and crisis, and present illustrative examples of real-world practices. In discussing these perspectives, we reflect on how we may better value, use, and exchange knowledge and practice in order to address current and future urban challenges.

Keywords: urban development; sustainability; urban resilience; crisis; flexibility; innovation; knowledge production

1. Introduction

The irreversible transition towards urban living entails complex challenges and vulnerabilities for citizens, civic authorities, and the management of global commons. By and large, these challenges are diverse and well known. Political and institutional fragmentation and rising inequality and austerity are changing the structural conditions for cities and citizens. Population growth and migration are challenging how cities are planned and governed, affecting their ability to provide basic services, housing, mobility, and infrastructure needs [1,2]. The growth and diversification of urban populations and the spatial expansion of cities strains the bureaucratic and technical capacities of how cities can plan and govern territories, provide services to citizens, and manage the competing needs and demands of growing urban constituencies [3,4]. Moreover, as cities become larger, more complex, diverse, and contested, addressing issues of inequality and exclusion and managing crime and security become more challenging [5].

Overarching these, environmental change and the disruptive nature of global health crises and pandemics also loom as specters threatening cities. Unmanaged and unregulated spatial and land-use plans contribute to urban areas becoming vulnerable to climate change impacts [6], and the sudden-onset and long-term impacts of climate change represent existential ecological threats [7,8]. Moreover, the COVID-19 pandemic has illustrated cities' political, economic, and social vulnerability caused by health emergencies that are unlikely to abate in the coming decades. All of these changes have fundamental implications for the resilience and vulnerability of societies and how cities are planned, governed, and lived in [9,10].

These brief examples are neither an exhaustive nor new set of challenges. Yet they are increasingly recognized and framed through a lens of multidimensional global crisis and systemic risk [11,12] with an explicitly urban characteristic [13,14]. Moreover, while rapidly urbanizing cities of the Global South are often pointed to as facing some of the most severe challenges or possessing less capacity to manage them [15–18], highly developed and urbanized countries also face increasingly fragile urban systems [19]. This has led to the widespread embrace of urban resilience perspectives [20], with cities becoming central to achieving sustainable development agendas. Yet the scale, pace, intensity, and complexity of challenges cities will face in the coming decades may render more established approaches to urban resilience and sustainability thinking insufficient. Pelling, for instance, has noted how common framings of resilience may tend to reproduce the status quo [21] rather than encourage transformative or transformational urban change that encompasses potentially radical shifts in structures, institutions, cultures, and practices in urban systems that may facilitate more resilient and sustainable cities [22–24]. While more transformational approaches to resilience are emerging in areas of reflexive governance [25–28] and urban experimentation and innovation [29], there are still mixed results regarding how more 'transformational' insights are integrated into broader urban policy and practice [30].

In light of this, this article offers a conceptual reflection with illustrative empirical examples on how urban crisis may shape how knowledge is produced, used, and shared to support urban resilience. This perspective is anchored in the idea that crises have the potential to represent periods of reflection and possibility that may necessitate or encourage new or novel practices to contend with social, political, economic, and environmental challenges. This is broadly informed by the idea of reflexivity and its considerations of how to support responsive change to address socio-ecological challenges while also maintaining a sufficient degree of systemic stability and integrity [31,32]. Here reflexivity is seen as "the capacity of an agent, structure, process or set of ideas to change in the light of reflection on its performance" [25] (p.942).

We contend that cities experiencing crises can be places where such 'reflexive' knowledge and practices may emerge. Our central assertion is that cities beset by fragility, crisis, and vulnerability—be it political, infrastructural, social, economic, or otherwise—can and do constitute important sites of knowledge production for urban resilience practices, and that they produce insights that we should pay greater attention to. This reflection is based on two premises. First, practices emerging in 'crisis cities' that are institutionally constrained or resource-limited may represent new or flexible models of resilience. Crisis can often represent a juncture where 'ideal' solutions are not possible, traditional approaches are no longer effective, or the consequences of delayed action are punitive. In these contexts, more transformative actions that are outside the scope of the 'conventional' may become permissible or necessary, opening space for new and potentially insightful knowledge and practices to emerge. Second, like other recent contributions [33], we maintain that ideas surrounding the production and use of knowledge for urban resilience must broaden to recognize and acknowledge alternative perspectives and practices. Doing so challenges dominant framings or established pathways of knowledge distribution of how urban resilience should be practiced or enacted [34] and where knowledge or expertise should emerge from [35–37]. Moreover, as cities across the globe will increasingly face vulnerabilities in the coming decades, it is imperative that we more thoughtfully consider how knowledge emerging from crises can inform how we act to address future crises in other contexts [38,39]. In light of this, we outline in Figure 1 a representation of how urban crisis may shape processes of knowledge production.

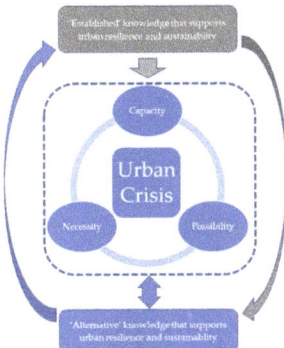

Figure 1. A conceptual representation of urban crisis and knowledge production.

In this representation, where "resilience continues to be mainly externally defined by expert knowledge from academia, international organization and governmental agencies" [40] (p. 257), 'established' forms of knowledge and practice tend to flow unidirectionally toward contexts of urban crisis. Yet as crisis impacts institutions, markets, and urban systems, these contexts are variously characterized by (i) constraints on resources and capacity; (ii) the necessity or urgency of action to address crisis; and (iii) a greater possibility for alternative actions [41]. Due to these conditions, established forms of knowledge that may support resilience may be resisted, ineffective, or inadequate to address the challenges at hand [42]. Yet, in parallel, alternative forms of knowledge iteratively emerge from local constellations of crisis as shaped by the nature of capacity, necessity, and possibility. Here, these alternative knowledges and practices are both grounded in local characteristics and influenced by established knowledge and practice and may constitute new or novel forms of resilience [43]. We argue that greater attention should be paid to how the dynamics of urban crisis shape such forms of new knowledge and practices to address vulnerabilities and how these could be integrated and shared to inform broader resilience and sustainability practices.

Having outlined the contours of our position, the article proceeds as follows. The next section briefly reviews selected insights from the sustainability, resilience, and crisis literatures and considers how crises may present opportunities for innovative or alternative knowledge and practice to emerge. Section 3 considers how cities in crisis may represent new sites of practice and their place in the processes of knowledge production and learning and anchor our reflections in illustrative examples from Medellín, Colombia, and Detroit, USA. Section 4 discusses broader implications and Section 5 concludes.

2. Crisis, Sustainability, and Resilience

The recognition of the challenges that cities face has shaped the evolving conception of 'urban crisis'. Emerging out of the confluence of racial tensions and neoliberal austerity in the US in the 1960s and 1970s, 'urban crisis' was deployed as a device to both describe the impacts of these forces and justify interventions to counteract them [44]. Yet the notion of urban crisis is diffuse and deployed to various ends. Often, crisis is viewed as an aberration from a prevailing 'normal', an interruption of the status quo, or a disruption of an otherwise 'acceptable' state of affairs. Here, Novalia and Malekpour note that crises are often framed as "special event(s) of exogenous origin punctuating the evolutionary dynamics of prevailing socio-technical or socio-ecological systems" [45] (p. 361). Others see crisis as a persistent feature of socio-technical systems under neo-liberal capitalism such that crisis is a chronic condition of societies today [46–48]. We consider 'urban crisis' to refer to contexts where the scale and/or magnitude of interconnected vulnerabilities that cities face present severe immediate or long-term challenges. The remainder of this section considers how crisis, sustainability, and resilience shape how we think about urban practice and the extent to which crisis may serve functions of "perpetuating the status quo, or, triggering systemic transformation" [45] (p. 361).

2.1. Averting Crisis: Sustainability and Resilience

Urban sustainability and urban resilience have become central to how we understand and contend with chronic and acute crises. While defined in a range of ways, urban sustainability and urban resilience can be viewed as emphasizing adaptability, flexibility, and the ability to respond to external shocks [49–51], and see cities as complex, interlinked and adaptive social, ecological, political, cultural, and economic systems that are prone to vulnerability in the face of new challenges [52–55]. Although both may be broadly considered to "understand system dynamics, enhance strategic competencies, and include diverse perspectives" [50] (p. 38), resilience and sustainability approaches also differ in important ways, with the boundary conditions of each concept often contested and critiqued [34,56–58].

Broadly, the resilience concept may be viewed as a more passive approach pursued with the purpose of understanding and responding to uncertainty, vulnerability, and the ability of systems to cope with shocks and crises. Alternatively, sustainability might be seen as a more active, deeper, adaptive endeavor devoted to the protection and maintenance of systems that provide social, economic, human, and ecological benefits [55,56]. One may consider resilience approaches as focusing on the process of systemic changes and practices, while sustainability approaches have a greater focus on the outcomes of such actions [50]. For this article, we consider urban resilience as the "ability for any urban system, with its inhabitants, to maintain continuity through all shocks and stresses" [59], and urban sustainability as the adaptive actions and processes that balance current and future ecological, economic and social interests "in response to changes within and beyond urban settlements" [60] (p. 213). By using these definitions, we view sustainability and resilience as related yet distinct concepts, but also concepts that should be interrogated regarding how they relate to contexts of urban crisis.

Therefore, a principal interest in this article is to consider how the knowledge and practices that emerge from urban crisis contexts are used to address and manage conditions of acute and chronic vulnerability. Here we focus on how crisis may inform how urban resilience and sustainability are conceptualized and practiced, rather than simply seeing them as being imposed from the outside. This interest is driven by urban resilience and sustainability thinking remaining encumbered by what we see as two particular challenges. First, there is a relative inflexibility of underlying urban systems [61] that can be seen as a structural barrier to implementing and scaling new practices. Infrastructure systems, institutional and governance structures, and other social and economic conditions can limit how cities alter how they respond, react, or engage with the challenges they face. For instance, Childers and co-authors highlight the problems related to inertia and the lack of flexibility that hinders change and may make cities more prone to vulnerability [62].

Second, and related to this structural inflexibility, there is a tendency to pursue a continuation of the status quo or incrementalism that reduces space for transformative or reflexive approaches [21,57,63]. Here, 'established' approaches and actions often emerge through the reproduction of existing 'expertise' and orthodoxy that may be inflexible or inadequate to meet the challenges at hand. These may also tend to "support particular types of state–society relations, construct particular kinds of at-risk subjects, and privilege technocratic solutions to disaster vulnerability" [64] (p. 1327). Over time, this incrementalism may see a fragmentation of the logic that underlies these approaches, such as the global or local situatedness that action should be taken [65]. Such challenges can be seen in the sustainable cities discourse [66,67], which broadly encompasses an approach to balance social and environmental concerns with urban growth in light of the ecological, social, political, and economic challenges that cities face [68]. In identifying how a narrower techno-economic focus is increasingly defining discussions around sustainable cities at the expense of justice and equity concerns, Hodson and Marvin note that "the sustainable city appears to be weakening as the dominant policy or research discourse of the future of the urban environments" [69] (p.9).

Given the inflexibility, incrementalism, and top-down focus inherent in many of the more 'traditional' resilience and sustainability approaches, the possibility of alternative knowledge that may emerge from conditions of crisis to support urban resilience is often precluded. This can hinder more reflexive or transformative knowledge production. McKinnon and Derickson note that the idea of resilience "privileges the restoration of existing systemic relations rather than their transformation" [57] (p. 262). This further speaks to the fact that operationalizing and implementing resilience approaches and practices is not 'power-neutral' and may, in many cases, overlook issues of environmental and human justice and equity. Meerow and others [70–72] have outlined, both theoretically and empirically, how the instrumental nature of the predominant conceptualizations of resilience can mask or exacerbate underlying or structural inequities and vulnerabilities, with similar critiques emerging in the sustainability literature [73]. Given this, there are increasing calls for more progressive approaches and operationalizations of the resilience concept that attend to issues of justice and equity in building resilience [34].

In briefly highlighting select aspects of discussions regarding urban resilience and sustainability as relates to the arguments in this paper, we note that while contested and, at times, overlapping concepts, they are both concerned with improving the ability of cities to avert or contend with systemic stresses. Both approaches, however, may tend towards inflexibility, inadequacy, or status quo thinking that undervalues novel or transformative action and may overlook issues of equity and justice. Despite this, the potential for shifts in power, agency, and justice is recognized in parts of the literature that emphasize how shocks to socio-ecological systems allow for reconfigurations and adaptations through cycles of growth and decline. Holling has outlined how following periods of growth and expansion in the 'front-loop', crises may engender a systemic reorganization and adaptation in the 'back-loop', which releases the potential for transformational action and response [74,75]. Viewing urban crises this way, as potential junctures for such 'systemic reorganization', opens a greater consideration of the forms of resilience emerging from contexts of crisis rather than simply as resilience thinking being applied to contexts of crisis. Here, by recognizing the types of knowledge and practices to address vulnerability that emerge from such contexts, so may we broaden how we conceptualize practices of sustainability and resilience.

2.2. Embracing' Crisis: Critical Crisis Theory and Broken World Thinking

Critical crisis scholarship broadly examines how crises may be able to catalyze new, innovative, or flexible practices. Similar to the back-loop discussed above, crises may challenge established political, social, and institutional practices, norms, and systems, promote the emergence of grassroots organizations and movements, and lead to reflections on what actions are possible, necessary, or legitimate [76]. For instance, the destructive nature of crises of capital has long been noted to create conditions for technological innovation and a return to growth, where the "politics in the wake of crises serves as a form of capitalist reconstruction delivering new opportunities" [77] (p. 38). Similarly, in discussions around resilience and disaster recovery, it has been suggested that "the radical potential of disaster lies in the experience of rupture that shifts the way individuals and communities see the world and the way society operates" [78] (p. 9). This notion of crisis as a 'critical juncture' has parallels with urban modes of living. The agglomeration effects in cities may enable new forms of action to emerge from "community interventions that are characterized by a desire to challenge the dominant norms and values of society and to experiment with different relationships and networks" [78] (p. 9). Explicitly noted by De Balanzó and Rodríguez-Planas, for instance, crisis was central to urban reorganization in Barcelona [79], where new interests and social movements came to challenge and negotiate prevailing or dominant urban practices in the 'back-loop phase' of the city's recent historical trajectories.

Such perspectives, thus, often frame crisis as an opportunity [77]. The destructive nature of crisis can have a reordering effect on societies by compelling or allowing for new constellations of practice to become necessary and/or possible and be a catalyst for social transformation. Works by Morin [80,81] suggest that crises reveal uncomfortable aspects of society and shine a light on dynamics that may require change, and while destructive, can also unleash transformational forces. Recently, Cordero and others noted that "crises are a reflexive moment for social actors to be able to put into question the norms and institutions that govern the present organization of society because those very conditions produce human suffering and become increasingly intolerable" [82] (p. 515).

These perspectives also parallel scholarship in the fields of media and technology on the failure and repair of infrastructure and hardware. Steven Jackson forwards the concept of Broken World Thinking, which argues that we see the use and re-use of material goods through the lens of collapse, decay, and failure. Rather than material goods being necessarily characterized by newness and optimal functionality, Jackson suggests that it is "erosion, breakdown, and decay, rather than novelty, growth, and progress" [83] (p. 221), which should be the starting points for thinking about physical products and systems and the ways that they can be used, applied, or enacted. Inherent here is a perspective that the rebuilding, repurposing, and reapplication of physical products is supported by and necessitated in contexts of failure or disrepair.

Extending this perspective from the realm of physical infrastructure and hardware to social systems, there are interesting parallels to draw when looking at situations of urban crisis. Here, in the context of the social, physical, institutional, and ecological systems of cities under stress, new practices can emerge that are able to overcome these challenges. In essence, to what extent does, or can, the breakdown of functional urban systems lead to the emergence of creative repurposing of the instruments of policy and civic action that lead to new, flexible constellations of practice and praxis in response to this systemic breakdown in 'crisis cities'? Jackson contends that we live in an "always-almost-falling-apart world" [83] (p. 222) where physical technologies and infrastructures are in an endless state of decay and disrepair, but also where there is a sense of wonder and appreciation at how lives are built and sustained around the restoration and persistence against the forces of disorder and breakdown. Thus, rather than a bleak vision of societies we live in, Jackson's view reiterates the agency in the face of collapse. It entails a promise of new beginnings as the world is in a "constant process of fixing and reinvention, reconfiguring and reassembling into new combinations and new possibilities". Here, the notion of repair does not simply entail patching together existing structures or institutions; it also fundamentally entails creativity, novelty, innovation, and transformation [83].

Reflecting on this, we draw a parallel to urban systems and the actors within them, considering how crises can invoke the need, and necessity, to re-evaluate, rethink and innovate in cities to encourage sustainable and resilient urban practices. Here, urban crises—be they sudden or slow onset—can present moments of innovation, novelty, and reflection that may initiate changes to mindsets, relationships, practices, policies, behaviors, material structures and urban systems that may lead to more just and sustainable urban outcomes. Thus, as crises may create conditions for flexible and novel solutions to emerge to address challenges facing cities, understanding such practices can provide valuable insights into how cities can adapt to and manage a future of unpredictable urban change.

3. Urban Crisis and the Production of Knowledge

This article calls for a deeper appreciation and understanding of cities experiencing crisis as sites of knowledge production and flexible practices that can support urban resilience. Here, we argue that important, novel, and often non-intuitive formal and informal solutions to urban challenges are arising out of urban vulnerability and complexity. Yet, as we note in the previous section vis-à-vis sustainability and resilience thinking, such practices that are nested in and emerge from specific socio-political contexts are often overlooked [84], despite the fact that these 'unexpected' cities are "making a virtue out of necessity have become world leaders in urban innovation" [85] (p. 337). Thus in 'crisis cities', knowledge and practice attuned to the constraints, necessities, and possibilities of local contexts can emerge in parallel, opposition to, or in concert with expert knowledge or 'best practice'. As Weichselgartner and Kelman observe, "decontextualized top-down knowledge on resilience offers a severely limited guide to operational practice, and may have considerably less purchase in problem-solving than pursuing co-designed bottom-up knowledge" [40]. As such, rather than a reliance on expert, external knowledge, actions to support resilience in the face of urban crises could rather see "urban systems ... re-imagined and re-designed by local actors with support from international organizations, not the reverse" [86] (p. 12).

Taking stock of this, it would be beneficial for scholars to consider in more detail how crisis may be constitutive in shaping new urban practices and alternative sites of knowledge production. In referring back to our conceptual representation in Section 1, we suggest that new knowledge and novel practices may emerge due to three conditions that characterize the experience of crisis.

First, crisis tends to limit the capacity for action available to address it. Financial, technical, geographic, political, or other resources may be acutely or persistently limited, which may constrain the range of ways that cities and citizens may 'ideally' seek to address a challenge. Moreover, established approaches may be less effective or incompatible with capacity-limited contexts. Instead, addressing crises in such instances may require flexible approaches to working through crisis in the absence of ideal or necessary capacities, encouraging new approaches in governing, organizing, and responding. This parallels with ideas of 'latent' social capital or capacity being activated to support adaptation activities, where underlying social bonds may reveal themselves through flexible forms of collective action, organization, or mobilization in response to socio-ecological vulnerability [87,88].

Second, crisis, with its negative impacts on societies, may tend to increase the necessity of action. Where more stable contexts may see more cautious, patient, or deliberate actions, crisis may engender an urgency that requires an acceleration of action. It may catalyze coalitions or interests around issues that may otherwise not be possible, force decision-making bottlenecks to be overcome, or shorten timelines for intervention where the alternative is to continue to suffer the (worsening) effects of crisis.

Finally, in addition to altering capacities in a way that may foster non-traditional actions and incentivize the necessity or urgency to act with purpose, crisis also opens the range of possibility for action. Crisis may represent moments where non-traditional courses of action can be pursued by states, when more traditional policy actions may prove ineffective or insufficient to address the challenges at hand [89,90]. Alternatively, manifestations of crisis may see once stable or impenetrable socio-political orders or organizing structures become contested or fragile and represent openings for new, often bottom-up forms of social innovation, resistance, collective action, or mobilization [91–93]. This idea also intersects with perspectives in other literature on moments of reorientation following shocks or crises. In the natural hazards and politics of disaster literature, periods of time following exogenous shocks or disasters are often conceptualized as a 'window of opportunity' to effect more radical forms of change [94,95]. Similarly, other literature considers the relevance of ecological concepts of 'disturbances' and their application to urban systems and how they may support the reflexive emergence of new or resilient practices [96].

Broadly then, crisis may be seen as a juncture where the societal or institutional constraints placed on particular sets of actions may soften or wane, and where the possibilities of new forms of action and the potentials for transformation are heightened. Moreover, an appreciation for urban crisis as a potential site of knowledge production can counteract or offer alternatives to predominating status quos. Here, attending to new sites and types of knowledge emerging from crisis challenges the centrality of formal, technocratic, 'best-practice' approaches to resilience. For instance, novel and effective solutions emerging from urban crises may not be top-down policy prescriptions but bottom-up or multi-stakeholder approaches that promote adaptive, just, and flexible approaches. This also contests the notion that cities encountering crisis are somehow 'failing' or are knowledge and capital 'collectors' rather than 'suppliers'. In particular, the literature on knowledge and policy exchange and diffusion, with few exceptions [97], are often rooted in the implicit assumption that knowledge flows from north to south, from 'successful' to 'strained' cities, or from 'formal' contexts to 'less formal' ones [98,99]. However, as Simone outlines in The City Yet to Come, by viewing Africa's metropolises as 'failing', we overlook opportunities to understand and capitalize on the myriad practices and structures in these cities that work under challenging conditions and that we may draw inspiration or lessons from [100]. In being open to the potential of new sites of knowledge production being shaped by the experience of crisis, we also then embrace the possibility that the flexible practices emerging in these places may transcend the contexts from where they came and have relevance for how we engage elsewhere with vulnerable and relatively inflexible urban systems [101,102].

Returning then to the title of this paper, what can we learn from urban crisis? At this stage, we have argued three main points. First, a series of interlinked, multidimensional challenges and crises will come to be constitutive of the urban condition in the future. Second, current approaches to sustainability and resilience thinking may lack the flexibility and reflexivity to manage these challenges and risk reproducing the conditions that permit the continuity of crisis. Third, flexible approaches to addressing vulnerability may emerge in contexts of urban crisis, and these may have implications for the production of knowledge for urban sustainability and resilience. In considering this further, we outline below a series of practices drawn from both the literature and our own primary research that shows real-world examples of the conceptual perspectives in this paper [41]. Rather than presenting a particular empirical argument, they should be seen as indicative cases that highlight the diverse ways in which new forms of organization and knowledge production are addressing urban challenges.

Practices

New, flexible urban practices are emerging in a number of areas, with many relating to citizens using creative means to overcome infrastructure, energy, or service delivery deficits. For instance, urban utility infrastructures are not necessarily ubiquitous [103], and particularly in the Global South, many creative solutions to this gap have emerged. Informal settlements often informally connect to water or electricity grids or establish connection and billing agreements with the state, as cases from Tanzania [104], India [105], and Bangladesh [106] show. While the creative repurposing of urban infrastructure is traditionally resisted by authorities and seen as 'anti-developmental', these actions also constitute sources of insight and innovation for infrastructural improvement.

Other cities have used creative governance approaches and planning reforms to address urban crises, exploring more participatory distributed systems in governance and for the urban environment. For example, Medellín, where we undertook qualitative fieldwork in 2018, is a well-known example of a city reimagining itself through a series of multi-stakeholder, participatory urban interventions over the past two decades. Broadly, the city has improved marginalized neighborhoods through holistic approaches that incorporate high-quality education, transport, utility infrastructure, and public space interventions [107]. Through its social urbanism approach, Medellin has been transformed from one of the world's most violent cities to a recognized hub of innovation and progressive urban practice and change [108,109]. The process was characterized by broad and interdisciplinary collaborations as well as a partnership between the city government and the public utility company Empresas Publicas de Medellín (EPM), exemplifying how architects, planners, engineers, and politicians used unique, innovative, and participatory approaches to reimagine a city that was fundamentally in crisis. As noted by a former mayor of Medellin, "I am certain the changes (policy and material interventions) in the city have made changes in the citizens and to citizenship . . . (To make these changes) you need something to bring together all these different sectors to work together. (You need) the unifying challenge and the leadership to unify" [110].

Other urban challenges that cut across issues of urban infrastructure, planning, governance, urban space, and the environment are seen in the decline of the post-industrial city. Detroit, Michigan, in the USA, for instance, where we undertook qualitative fieldwork in 2019, was an industrial powerhouse in the mid-twentieth century, yet the loss of its manufacturing base caused a remarkable population decline and middle-class flight to the suburbs in the following decades. As the municipal tax base declined, exacerbated by the Global Financial Crisis, which decimated homeownership through bankruptcies and foreclosures, the management of the physical city and delivery of services became severely constrained. The city was ultimately governed into demise, declaring bankruptcy in 2013, with the former mayor Kwame Kilpatrick imprisoned for corruption and financial mismanagement [111,112]. Yet despite resource constraints, extensive poverty, land and home vacancy, and blight in the city, the past decade has seen a resurgence; new community initiatives, social movements, innovative planning approaches, and social entrepreneurship have emerged. These new practices include alternate forms of community-driven governance, urban greening and farming, and new forms of land tenure following the city's real estate collapse [113], representing diverse strategies for urban reinvention and reconfiguration of social, governance, and infrastructure systems [114]. With some considering Detroit's crisis as an opportunity for reinvention and renewal [115,116], the enthusiasm around the city has seen Detroit be called "the most exciting city in America", and that it has turned its "end of days into a laboratory of the future" [117].

While far from exhaustive, these brief illustrative examples outline the potential for urban challenges and multidimensional crises to shape the conditions for—and often enable—new forms of civic and state interventions that address complexity and vulnerability. As we have noted elsewhere, it calls on us to re-evaluate the mindsets, actors, behaviors, relationships, structures, and resources needed to allow these to flourish and pursue more inclusive, flexible urban transformations [41].

4. Discussion

This paper argues that novel, flexible urban practices that emerge out of conditions of crisis in cities across the world need to be better understood, both in how and what type of knowledge is produced and how knowledge is used. As we have outlined, established knowledge and practice may be unsuited to contexts of urban crisis, but also that the nature of capacity, necessity, and the possibility of action during crisis may produce new practices, solutions and knowledge. Paying greater attention to these may benefit broader understandings of how we can support urban resilience in different ways.

Regarding the production of knowledge, at the outset, we should interrogate notions of what is regarded as 'best practice' or 'innovation'. For instance, the central 'narratives' or common 'best practices' related to urban resilience and sustainability often view knowledge as produced in western, formal, or 'stable' contexts and transferred to southern, informal, or 'crisis' contexts [118]. Yet this is often done without sufficient appreciation for grassroots or subaltern forms of action that may value alternative types of knowledge or, more concretely, center issues of participation, equity, and justice in urban resilience and sustainability practices [119]. Indeed, in assessing north–south knowledge transfer and city to city cooperation, Mayer and Long note that these initiatives were "more likely to support than challenge entrenched practices which can weaken sustainable development governance" [120] (p. 1). For instance, technocratic approaches that often drive urban sustainability and resilience agendas [121] are by and large beholden to prevailing 'best practice' governance logics and status quos that may themselves create conditions for crisis [122,123]. Moreover, a focus on 'best practice' can reinforce a unidirectionality of knowledge flows without appreciation for "translocal geographies of knowledge production and circulation" [124] (p. 10). This may preclude certain 'alternative' or bottom-up practices from taking root or occluding them from the toolkit of possible responses [125,126]. Here, May has observed a "culture of expertise that is at odds with democracy through a separation between the forms of justification it deploys and the contexts of its application... in which models and ideas for urban development circulate without sensitivity to context" [127] (p. 2189). Countering this requires a deeper focus on the knowledge and practice emerging in new places or from new actors that may be outside the scope of what is regarded as being 'typical', 'expert', or 'accepted'. May suggests this requires "a movement away from these narrowly constituted forms of knowledge production and reception to provide a responsible politics through a more open and inclusive approach to urban development" [127] (p. 2189). This sentiment is echoed in calls to understand better the nature of crisis in order to resist and challenge prevailing discourses and move towards more transformative urban politics and practice [128].

In uncovering and supporting a diversity of flexible practices, voices, and agencies of urban stakeholders, notions of urban experimentation have arisen. Discussing urban laboratories, Karvonen and van Heur note, "these spaces of innovation and change provide a designated space for experimentation where new ideas can be designed, implemented, measured and, if successful, scaled up and transferred to other locales" [129] (p. 11). Yet here, there is a need to be attentive to the conditions under which such experimental or innovative approaches emerge and how they consider equity, justice, and agency in these processes [130]. While crises create the potential for new forms of knowledge and practice to be produced through innovative or experimental actions, discourses and constructions of crisis can likewise be subjugated to existing structures of power or justify exclusionary urban interventions [131,132]. Thus, while some urban laboratories "make a genuine attempt to cultivate emancipatory forms of change that could have widespread implications on urban life in the twenty-first century and beyond", others "simply employ the notions of 'laboratory' and 'experiment' as a rhetorical strategy to further consolidate and reinforce existing patterns of urban development" [129] (p. 11).

More recently, the COVID pandemic has brought into sharp relief the inherent fragility in our urban systems, prompting new considerations about governance innovation and urban experimentation [133,134]. It has also shown the relevance of mobilizing and sharing knowledge to contend with emerging urban vulnerabilities [135]. It has rightly been noted that "COVID-19 has magnified the deficiencies of how we manage our cities but has also given us a unique chance to rethink, replan, and redesign" [134] (p. 318). It is thus possible to view this period as a critical juncture or moment of reflection regarding our current trajectories of urban development, sustainability, and resilience. It suggests a reconsideration of how we think about crisis, how the knowledge and practices to contend with crisis are produced, and how such 'alternative' knowledge is shared, used, and integrated with 'established' knowledge. As the urban condition of the 21st century will require the management of and response to a series of multidimensional challenges, so too should we re-evaluate what crisis, care, resilience, and sustainable practice can and could be.

5. Conclusions

We are witnessing a cleavage where many of the foundational aspects of our political, economic, ecological, and social systems are slowly being revealed as both acutely inflexible and inherently fragile [136]. Despite the emergence of resilience and sustainability practices in the past 30 years, cities remain largely underprepared for the challenges they will face in the coming decades. As Roitman notes, "crisis characterize(s) the world in which we act", but also rightly identifies how narratives of crisis can instrumentally enable certain responses to the exclusion of others [137] (pp. 73–74). Dissecting these narratives, understanding varied experiences of crisis, acknowledging the novel responses and practices that support resilience, justice, and equity in the city, and discerning what and how we can learn from these experiences have become imperative for urban research and practice.

This article has presented a modest conceptual reflection on the future of interconnected challenges in urban areas. We have offered that the dynamics of capacity, necessity, and possibility during urban crises may shape new forms of knowledge and practice that should be more systematically examined. In being more attentive to the new forms of urban policy, social organization, consensus building, and alternative practices that can emerge in places characterized by crisis and vulnerability, we are asked to reconsider the nature of knowledge production and knowledge sharing to support urban resilience. Here, seeing cities through a lens of crisis, flexibility and learning may reveal new knowledge pathways or entry points to address vulnerability and catalyze learning and engagement within and across cities. As crisis is becoming a defining condition of 21st-century cities, it should also be integral in influencing how we should respond. Future research focusing on identifying how cities and citizens, particularly in the Global South, have experienced and managed crisis, processes by which resilient practices emerge and are sustained, and how these may be scaled or transferred to other contexts would be fruitful forward agendas for scholarship.

Finally, while we must contend with the potential of crisis as a 'new normal', our collective pursuits should perhaps be focused on what Alarouf has elegantly termed our 'forgotten normal', and its expressions of community, social justice, respect for ecological sustainability, and people-based places and spaces. As he notes, there is "nothing more profound than times of crisis to inspire communities to create better existential positions" [138] (p. 169). It is prudent then that we begin to value the knowledge produced in cities that are contending with crisis and consider more thoroughly how we can use this knowledge to improve the prospects for a just and equitable urban future.

Author Contributions: Conceptualization, K.H., H.C.G., L.H. and S.A.P.; methodology, K.H., H.C.G., L.H. and S.A.P.; investigation, K.H., H.C.G., L.H. and S.A.P.; writing—original draft preparation, K.H., H.C.G., L.H. and S.A.P.; writing—review and editing, K.H., H.C.G., L.H. and S.A.P. All authors have read and agreed to the published version of the manuscript.

Funding: This research was funded by the Research Council of Norway, grant number 259906. The APC was funded by SINTEF, Trondheim, Norway.

Institutional Review Board Statement: Not applicable.

Informed Consent Statement: Not applicable.

Data Availability Statement: Not applicable.

Conflicts of Interest: The authors declare no conflict of interest.

References

1. UNHABITAT. *Introduction to ECOSOC Humanitarian Affairs Segment 2016: Urban Crises and the New Urban Agenda Organized by the United Kingdom Permanent Mission to the United Nations & UN Habitat, Wednesday, 29 June*; UNHABITAT: Nairobi, Kenya, 2016.
2. UNHABITAT. *World Cities Report 2016: Urbanization and Development. Emerging Futures*; UNHABITAT: Nairobi, Kenya, 2016.
3. Fainstein, S. *The Just City*; Cornell Press: New York, NY, USA, 2010.
4. Rotberg, R. *When States Fail: Causes and Consequences*; Princeton University Press: Princeton, NJ, USA, 2010.
5. Muggah, R. *Researching the Urban Dilemma: Urbanization, Poverty and Violence*; IDRC: Ottawa, ON, Canada, 2012.
6. Morita, A. Multispecies Infrastructure: Infrastructural Inversion and Involuntary Entanglements in the Chao Phraya Delta, Thailand. *Ethnos* **2016**, *82*, 738–757. [CrossRef]
7. Bulkeley, H. *Cities and Climate Change*; Routledge: London, UK, 2013.
8. Rosenzweig, C.; Solecki, W.D.; Hammer, S.A.; Mehrotra, S. *Climate Change and Cities: First Assessment Report of the Urban Climate Change Research Network*; Cambridge University Press: Cambridge, UK, 2011.
9. Fenton, P.; Gustafsson, S. Moving from high-level words to local action—Governance for urban sustainability in municipalities. *Curr. Opin. Environ. Sustain.* **2017**, *26–27*, 129–133. [CrossRef]
10. Bulkeley, H.; Broto, V.C. Government by experiment? Global cities and the governing of climate change. *Trans. Inst. Br. Geogr.* **2012**, *38*, 361–375. [CrossRef]
11. Ehrlich, P.R.; Ehrlich, A.H. Can a collapse of global civilization be avoided? *Proc. R. Soc. B Boil. Sci.* **2013**, *280*, 20122845. [CrossRef]
12. Homer-Dixon, T.; Walker, B.; Biggs, R.; Crépin, A.-S.; Folke, C.; Lambin, E.F.; Peterson, G.D.; Rockström, J.; Scheffer, M.; Steffen, W.; et al. Synchronous failure: The emerging causal architecture of global crisis. *Ecol. Soc.* **2015**, *20*, 6. [CrossRef]
13. Acuto, M.; Larcom, S.; Keil, R.; Ghojeh, M.; Lindsay, T.; Camponeschi, C.; Parnell, S. Seeing COVID-19 through an urban lens. *Nat. Sustain.* **2020**, *3*, 977–978. [CrossRef]
14. Elmqvist, T.; Andersson, E.; McPhearson, T.; Bai, X.; Bettencourt, L.; Brondizio, E.; Van Der Leeuw, S. Urbanization in and for the Anthropocene. *Npj. Urban Sustain.* **2021**, *1*, 1–6. [CrossRef]
15. Pieterse, E. *City Futures: Confronting the Crisis of Urban Development*; Zed Books: London, UK, 2008.

16. Davis, M. *Planet of Slums*; Verso: New York, NY, USA, 2006.
17. Cobbinah, P.B.; Erdiaw-Kwasie, M.O.; Amoateng, P. Africa's urbanisation: Implications for sustainable development. *Cities* **2015**, *47*, 62–72. [CrossRef]
18. Hove, M.; Ngwerume, E.T.; Muchemwa, C. The Urban Crisis in Sub-Saharan Africa: A Threat to Human Security and Sustainable Development. *Stab. Int. J. Secur. Dev.* **2013**, *2*, 7. [CrossRef]
19. Selby, J.D.; Desouza, K. Fragile cities in the developed world: A conceptual framework. *Cities* **2019**, *91*, 180–192. [CrossRef]
20. Ribeiro, P.J.G.; Pena Jardim Gonçalves, L.A. Urban resilience: A conceptual framework. *Sustain. Cities Soc.* **2019**, *50*, 101625. [CrossRef]
21. Pelling, M. *Adaptation to Climate Change: From Resilience to Transformation*; Routledge: London, UK, 2011.
22. Hölscher, K.; Frantzeskaki, N. Perspectives on urban transformation research: Transformations in, of, and by cities. *Urban Transform.* **2021**, *3*, 1–14. [CrossRef]
23. Wolfram, M.; Borgström, S.; Farelly, M. Urban transformative capacity: From concept to practice. *Ambio* **2019**, *48*, 437–448. [CrossRef]
24. Evans, J.; Vácha, T.; Kok, H.; Watson, K. How Cities Learn: From Experimentation to Transformation. *Urban Plan.* **2021**, *6*, 171–182. [CrossRef]
25. Dryzek, J.S. Institutions for the Anthropocene: Governance in a Changing Earth System. *Br. J. Politi Sci.* **2016**, *46*, 937–956. [CrossRef]
26. Ferrari, M. Reflexive Governance for Infrastructure Resilience and Sustainability. *Sustainability* **2020**, *12*, 10224. [CrossRef]
27. Voß, J.-P.; Borneman, B. The politics of reflexive governance: Challenges for designing adaptive management and transition management. *Ecol. Soc.* **2011**, *16*, 9. [CrossRef]
28. Feindt, P.H.; Weiland, S. Reflexive governance: Exploring the concept and assessing its critical potential for sustainable development. Introduction to the special issue. *J. Environ. Policy Plan.* **2018**, *20*, 661–674. [CrossRef]
29. Bulkeley, H.; Marvin, S.; Palgan, Y.V.; McCormick, K.; Breitfuss-Loidl, M.; Mai, L.; Von Wirth, T.; Frantzeskaki, N. Urban living laboratories: Conducting the experimental city? *Eur. Urban Reg. Stud.* **2019**, *26*, 317–335. [CrossRef]
30. Scholl, C.; de Kraker, J. Urban Planning by Experiment: Practices, Outcomes, and Impacts. *Urban Plan.* **2021**, *6*, 156–160. [CrossRef]
31. Pickering, J. Ecological reflexivity: Characterising an elusive virtue for governance in the Anthropocene. *Environ. Politics* **2018**, *28*, 1145–1166. [CrossRef]
32. Dryzek, J.; Pickering, J. *The Politics of the Anthropocene*; Oxford University Press: Oxford, UK, 2018.
33. Feagan, M.; Matsler, M.; Meerow, S.; Muñoz-Erickson, T.A.; Hobbins, R.; Gim, C.; Miller, C.A. Redesigning knowledge systems for urban resilience. *Environ. Sci. Policy* **2019**, *101*, 358–363. [CrossRef]
34. Vale, L.J. The politics of resilient cities: Whose resilience and whose city? *Build. Res. Inf.* **2014**, *42*, 191–201. [CrossRef]
35. Goldstein, B.E.; Wessells, A.T.; Lejano, R.; Butler, W. Narrating Resilience: Transforming Urban Systems Through Collaborative Storytelling. *Urban Stud.* **2013**, *52*, 1285–1303. [CrossRef]
36. Borie, M.; Pelling, M.; Ziervogel, G.; Hyams, K. Mapping narratives of urban resilience in the global south. *Glob. Environ. Chang.* **2019**, *54*, 203–213. [CrossRef]
37. Wijsman, K.; Feagan, M. Rethinking knowledge systems for urban resilience: Feminist and decolonial contributions to just transformations. *Environ. Sci. Policy* **2019**, *98*, 70–76. [CrossRef]
38. Boin, A.; McConnell, A.; Hart, P.T. *Governing after Crisis: The Politics of Investigation, Accountability and Learning*; Oxford University Press: Oxford, UK, 2008.
39. Deverell, E. Crises as learning triggers: Exploring a conceptual framework of crisis-induced learning. *J. Cont. Crisis Manag.* **2009**, *17*, 179–188. [CrossRef]
40. Weichselgartner, J.; Kelman, I. Geographies of resilience: Challenges and opportunities of a descriptive concept. *Prog. Human Geogr.* **2015**, *39*, 249–267. [CrossRef]
41. Suyama, B.; Amaro, G.L.; Geirbo, H.C.; Harboe, L.; Hoelscher, K.; Martins, D.; Petersen, S.A. Learning Flexibility: Pathways to Urban Transformation. 2021. Available online: www.learningflexibility.com (accessed on 29 October 2021).
42. Shamsuddin, S. Resilience resistance: The challenges and implications of urban resilience implementation. *Cities* **2020**, *103*, 102763. [CrossRef]
43. DeVerteuil, G.; Golubchikov, O.; Sheridan, Z. Disaster and the lived politics of the resilient city. *Geoforum* **2021**, *125*, 78–86. [CrossRef]
44. Weaver, T. Urban crisis: The genealogy of a concept. *Urban Stud.* **2016**, *54*, 2039–2055. [CrossRef]
45. Novalia, W.; Malekpour, S. Theorising the role of crisis for transformative adaptation. *Environ. Sci. Policy* **2020**, *112*, 361–370. [CrossRef]
46. Gotham, K.F.; Greenberg, M. *Crisis Cities: Disaster and Redevelopment in New York and New Orleans*; Oxford University Press: Oxford, UK, 2014.
47. Perrow, C. *Normal Accidents: Living with High-Risk Technologies—Updated Edition*; Princeton University Press: Princeton, NJ, USA, 2011.
48. Vigh, H.E. Crisis and Chronicity: Anthropological Perspectives on Continuous Conflict and Decline. *Ethnos* **2008**, *73*, 5–24. [CrossRef]
49. Jabareen, Y. Planning the resilient city: Concepts and strategies for coping with climate change and environmental risk. *Cities* **2013**, *31*, 220–229. [CrossRef]

50. Redman, C.L. Should Sustainability and Resilience Be Combined or Remain Distinct Pursuits? *Ecol. Soc.* **2014**, *19*, 37. [CrossRef]
51. Sanchez, A.X.; Van der Heijden, J.; Osmond, P. The city politics of an urban age: Urban resilience conceptualisations and policies. *Palgrave Commun.* **2018**, *4*, 25. [CrossRef]
52. Mehmood, A. Of resilient places: Planning for urban resilience. *Eur. Plan. Stud.* **2016**, *24*, 407–419. [CrossRef]
53. Coaffee, J. Protecting vulnerable cities: The UK's resilience response to defending everyday urban infrastructure. *Int. Aff.* **2010**, *86*, 939–954. [CrossRef]
54. Wilkinson, C. Social-ecological resilience: Insights and issues for planning theory. *Plan. Theory* **2012**, *11*, 148–169. [CrossRef]
55. Pickett, S.T.; Belt, K.T.; Galvin, M.F.; Groffman, P.; Grove, J.M.; Outen, D.C.; Pouyat, R.V.; Stack, W.P.; Cadenasso, M.L. Watersheds in Baltimore, Maryland: Understanding and Application of Integrated Ecological and Social Processes. *J. Contemp. Water Res. Educ.* **2009**, *136*, 44–55. [CrossRef]
56. Zhang, X.; Li, H. Urban resilience and urban sustainability: What we know and what do not know? *Cities* **2018**, *72*, 141–148. [CrossRef]
57. MacKinnon, D.; Derickson, K.D. From resilience to resourcefulness: A critique of resilience policy and activism. *Prog. Human Geogr.* **2013**, *37*, 253–270. [CrossRef]
58. Rogov, M.; Rozenblat, C. Urban Resilience Discourse Analysis: Towards a Multi-Level Approach to Cities. *Sustainability* **2018**, *10*, 4431. [CrossRef]
59. UNHABITAT. Available online: https://urbanresiliencehub.org/what-is-urban-resilience/ (accessed on 25 September 2021).
60. Wu, J. Urban ecology and sustainability: The state-of-the-science and future directions. *Landsc. Urban Plan.* **2014**, *125*, 209–221. [CrossRef]
61. Roggema, R. Towards Enhanced Resilience in City Design: A Proposition. *Land* **2014**, *3*, 460–481. [CrossRef]
62. Childers, D.L.; Cadenasso, M.L.; Grove, J.M.; Marshall, V.; McGrath, B.; Pickett, S.T.A. An Ecology for Cities: A Transformational Nexus of Design and Ecology to Advance Climate Change Resilience and Urban Sustainability. *Sustainability* **2015**, *7*, 3774–3791. [CrossRef]
63. Brown, K. Policy discourses of resilience. In *Climate Change and the Crisis of Capitalism: A Chance to Reclaim Self, Society and Nature*; Pelling, M., Manuel-Navarrete, D., Redclift, M., Eds.; Routledge: Oxon, UK, 2012.
64. Tierney, K. Resilience and the neoliberal project: Discourses, critiques, practices—And Katrina. *Am. Behav. Sci.* **2015**, *59*, 1327–1342. [CrossRef]
65. Hodson, M.; Marvin, S. *After Sustainable Cities?* Routledge: London, UK, 2014.
66. Williams, K. Sustainable cities: Research and practice challenges. *Int. J. Urban Sustain. Dev.* **2010**, *1*, 128–132. [CrossRef]
67. Flint, J.; Raco, M. *The Future of Sustainable Cities Critical Reflections*; Bristol University Press: Bristol, UK, 2011.
68. Campbell, S.D. Green cities, growing cities, just cities? Urban planning and the contradictions of sustainable development. *J. Am. Plan. Assoc.* **1996**, *62*, 296–312. [CrossRef]
69. Hodson, M.; Marvin, S. Intensifying or transforming sustainable cities? Fragmented logics of urban environmentalism. *Local Environ.* **2017**, *22*, 8–22.
70. Meerow, S.; Pajouhesh, P.; Miller, T.R. Social equity in urban resilience planning. *Local Environ.* **2019**, *24*, 793–808. [CrossRef]
71. Meerow, S.; Newell., J.P. Urban Resilience for Whom, What, When, Where, and Why? *Urban Geogr.* **2019**, *40*, 309–329. [CrossRef]
72. Meerow, S. Double exposure, infrastructure planning, and urban climate resilience in coastal megacities: A case study of Manila. *Environ. Plan. A Econ. Space* **2017**, *49*, 2649–2672. [CrossRef]
73. Agyeman, J.; Evans, T. Toward Just Sustainability in Urban Communities: Building Equity Rights with Sustainable Solutions. *Ann. Am. Acad. Political Soc. Sci.* **2003**, *590*, 35–53. [CrossRef]
74. Holling, C.S. The resilience of terrestrial ecosystems: Local surprise and global change. In *Sustainable Development of the Biosphere*; Clarck, W.C., Munn, R.E., Eds.; Cambridge University Press: Cambridge, UK, 1986; pp. 292–317.
75. Gunderson, L.H.; Holling, C.S. *Panarchy: Understanding Transformations in Human and Natural Systems*; Island: Washington, DC, USA, 2002.
76. Cordero, R. *Crisis and Critique: On the Fragile Foundations of Social Life*; Routledge: London, UK, 2016.
77. Stephanides, P. Crisis as Opportunity? An Ethnographic Case-Study of the Post-Capitalist Possibilities of Crisis Community Currency Movements. Ph.D. Thesis, University of East Anglia, Norwich, UK, 2017.
78. Cretney, R.M. Towards a critical geography of disaster recovery politics: Perspectives on crisis and hope. *Geogr. Compass* **2017**, *11*, e12302. [CrossRef]
79. De Balanzó, R.; Rodríguez-Planas, N. Crisis and reorganization in urban dynamics. *Ecol. Soc.* **2018**, *23*, 1–19. [CrossRef]
80. Morin, E. Pour une crisologie. *Communications* **1976**, *25*, 149–163. [CrossRef]
81. Morin, E. For a Crisology. *Indust. Environ. Crisis Q.* **1993**, *7*, 5–21. [CrossRef]
82. Cordero, R.; Mascareño, A.; Chernilo, D. On the reflexivity of crises: Lessons from critical theory and systems theory. *Eur. J. Soc. Theory* **2017**, *20*, 511–530. [CrossRef]
83. Jackson, S.J. Rethinking repair. In *Media Technologies: Essays on Communication, Materiality and Society*; Gillespie, T., Boczkowski, P.J., Foot, K.A., Eds.; MIT Press: Cambridge, UK, 2014; pp. 221–239.
84. Cote, M.; Nightingale, A.J. Resilience thinking meets social theory: Situating social change in socio-ecological systems (SES) research. *Prog. Human Geogr.* **2012**, *36*, 475–489. [CrossRef]
85. Hall, P.; Pfeiffer, U. *Urban Future 21: A Global Agenda for Twenty-First Century Cities*; Routledge: London, UK, 2000.

86. Sitko, P.; Massella, A. *Building Urban Resilience in the Face of Crisis: A Focus on People and Systems*; Global Alliance for Urban Crisis: Geneva, Switzerland, 2019.
87. Pelling, M.; High, C. Understanding adaptation: What can social capital offer assessments of adaptive capacity? *Glob. Environ. Chang.* **2005**, *15*, 308–319. [CrossRef]
88. Adger, W.N. Social Capital, Collective Action, and Adaptation to Climate Change. *Econ. Geogr.* **2009**, *79*, 387–404. [CrossRef]
89. Hoelscher, K.; Nussio, E. Understanding Unlikely Successes in Urban Violence Reduction. *Urban Stud.* **2016**, *53*, 2397–2416. [CrossRef]
90. Hoelscher, K. Institutional Reform and Violence Reduction in Pernambuco, Brazil. *J. Lat. Am. Stud.* **2017**, *49*, 855–884. [CrossRef]
91. Blanco, I.; León, M. Social innovation, reciprocity and contentious politics: Facing the socio-urban crisis in Ciutat Meridiana, Barcelona. *Urban Stud.* **2017**, *54*, 2172–2188. [CrossRef]
92. Widyaningsih, A.; Van den Broeck, P. Social innovation in times of flood and eviction crisis: The making and un-making of homes in the Ciliwung riverbank, Jakarta. *Singap. J. Trop. Geogr.* **2021**, *42*, 325–345. [CrossRef]
93. Seyfang, G.; Smith, A. Grassroots innovations for sustainable development: Towards a new research and policy agenda. *Environ. Politics* **2007**, *16*, 584–603. [CrossRef]
94. Birkmann, J.; Buckle, P.; Jaeger, J.; Pelling, M.; Setiadi, N.; Garschagen, M.; Fernando, N.; Kropp, J. Extreme events and disasters: A window of opportunity for change? Analysis of organizational, institutional and political changes, formal and informal responses after mega-disasters. *Nat. Hazards* **2010**, *55*, 637–655. [CrossRef]
95. Brundiers, K.; Eakin, H.C. Leveraging Post-Disaster Windows of Opportunities for Change towards Sustainability: A Framework. *Sustainability* **2018**, *10*, 1390. [CrossRef]
96. Grimm, N.B.; Pickett, S.; Hale, R.L.; Cadenasso, M.L. Does the ecological concept of disturbance have utility in urban social–ecological–technological systems? *Ecosyst. Health Sustain.* **2017**, *3*, e01255. [CrossRef]
97. Sanyal, B. Knowledge transfer from poor to rich cities: A new turn of events. *Cities* **1990**, *7*, 31–36. [CrossRef]
98. Clarke, N. Actually existing comparative urbanism: Imitation and cosmopolitanism in North-South inter-urban partnerships. *Urban Geogr.* **2012**, *33*, 796–815. [CrossRef]
99. Mocca, E. All cities are equal, but some are more equal than others. Policy mobility and asymmetric relations in inter-urban networks for sustainability. *Int. J. Urban Sustain. Dev.* **2018**, *10*, 139–153. [CrossRef]
100. Simone, A. *For the City yet to Come: Changing African Life in Four Cities*; Duke University Press: Durham, NC, USA, 2004.
101. Palmer, M.; Kramer, J.G.; Boyd, J.; Hawthorne, D. Practices for facilitating interdisciplinary synthetic research: The National Socio-Environmental Synthesis Center (SESYNC). *Curr. Opin. Environ. Sustain.* **2016**, *19*, 111–122. [CrossRef]
102. Grove, J.M.; Childers, D.L.; Galvin, M.; Hines, S.; Muñoz-Erickson, T.; Svendsen, E.S. Linking science and decision making to promote an ecology for the city: Practices and opportunities. *Ecosyst. Health Sustain.* **2016**, *2*, e01239. [CrossRef]
103. Graham, S.; Marvin, S. *Splintering Urbanism: Networked Infrastructures, Technological Mobilities and the Urban Condition*; Taylor & Francis: Abingdon, UK, 2001.
104. Winther, T. *The Impact of Electricity: Development, Desires and Dilemmas*; Berghahn Books: New York, NY, USA, 2008.
105. Anand, N. Municipal disconnect: On abject water and its urban infrastructures. *Ethnography* **2012**, *13*, 487–509. [CrossRef]
106. Ahmed, S.I.; Min, N.J.; Jackson, S.J. Residual Mobilities: Infrastructural Displacement and Post-Colonial Computing in Bangladesh. In Proceedings of the 33rd Annual ACM Conference on Human Factors in Computing Systems, Seoul, Korea, 18–23 April 2015; pp. 437–446.
107. Schwab, E.; Aponte, G. Small Scale—Big Impact? *Medellín's Integral Urban Projects. Topos* **2013**, *84*, 36.
108. Maclean, K. *Social Urbanism and the Politics of Violence: The Medellín Miracle*; Palgrave Macmillan: London, UK, 2015.
109. Varela Barrios, E. Expansion Strategies and Management Methods in Empresas Públicas de Medellín, EPM. *Estud. Políticos* **2010**, *36*, 141–165.
110. Interview, August 2018, Medellin, unpublished. 20 August.
111. Eisinger, P. Is Detroit Dead? *J. Urban Aff.* **2014**, *36*, 1–12. [CrossRef]
112. Neill, W.J.V. Carry on Shrinking?: The Bankruptcy of Urban Policy in Detroit. *Plan. Pr. Res.* **2014**, *30*, 1–14. [CrossRef]
113. Safransky, S. Greening the urban frontier: Race, property, and resettlement in Detroit. *Geoforum* **2014**, *56*, 237–248. [CrossRef]
114. Gallagher, J. *Revolution Detroit: Strategies for Urban Reinvention*; Wayne State University Press: Detroit, MI, USA, 2013.
115. Ferris, J.M.; Hopkins, E.M. Urban Crisis as Opportunity. *Stanf. Soc. Innov. Rev.* **2016**, *15*, A2–A5. [CrossRef]
116. Gallagher, J. *Reimagining Detroit: Opportunities for Redefining an American City*; Wayne State University Press: Detroit, MI, USA, 2010.
117. Binelli, M. *Detroit City is the Place to Be: The Afterlife of an American Metropolis*; Macmillan: New York, NY, USA, 2013.
118. Nagendra, H.; Bai, X.; Brondizio, E.; Lwasa, S. The urban south and the predicament of global sustainability. *Nat. Sustain.* **2018**, *1*, 341–349. [CrossRef]
119. du Toit, M.J.; Shackleton, C.M.; Cilliers, S.S.; Davoren, E. Advancing Urban Ecology in the Global South: Emerging Themes and Future Research Directions. *Cities Nat.* **2021**, *433*, 433–461. [CrossRef]
120. Mayer, L.; Long, L.A.N. Can city-to-city cooperation facilitate sustainable development governance in the Global South? Lessons gleaned from seven North-South partnerships in Latin America. *Int. J. Urban Sustain. Dev.* **2021**, *13*, 174–186. [CrossRef]
121. Zebrowski, C. Acting local, thinking global: Globalizing resilience through 100 Resilient Cities. New Perspectives. *Interdiscip. J. Cent. East Eur. Politics Int. Relat.* **2020**, *28*, 71–88. [CrossRef]
122. Gunder, M. Sustainability: Planning's saving grace or road to perdition? *J. Plan. Educ. Res.* **2006**, *26*, 208–221. [CrossRef]

23. Brown, T. Sustainability as Empty Signifier: Its Rise, Fall, and Radical Potential. *Antipode* **2016**, *48*, 115–133. [CrossRef]
24. Frediani, A.A.; Cociña, C.; Acuto, M. *Translating Knowledge for Urban Equality: Alternative Geographies for Encounters between Planning Research and Practice KNOW Working Paper No. 2*; KNOW: London, UK, 2019.
25. Weaver, T.P. Charting Change in the City: Urban Political Orders and Urban Political Development. *Urban Aff. Rev.* **2021**. [CrossRef]
26. Weaver, T.P. By design or by default: Varieties of neoliberal urban development. *Urban Aff. Rev.* **2018**, *54*, 234–266. [CrossRef]
27. May, T. Urban crisis: Bonfire of vanities to find opportunities in the ashes. *Urban Stud.* **2017**, *54*, 2189–2198. [CrossRef]
28. Bayırbağ, M.K.; Davies, J.S.; Münch, S. Interrogating urban crisis: Cities in the governance and contestation of austerity. *Urban Stud.* **2017**, *54*, 2023–2038. [CrossRef]
29. Karvonen, A.; Van Heur, B. Urban laboratories: Experiments in reworking cities. *Int. J. Urban Reg. Res.* **2014**, *38*, 379–392. [CrossRef]
30. Caprotti, F.; Cowley, R. Interrogating urban experiments. *Urban Geogr.* **2017**, *38*, 1441–1450. [CrossRef]
31. Oosterlynck, S.; González, S. 'Don't waste a crisis': Opening up the city yet again for neoliberal experimentation. *Int. J. Urban Reg. Res.* **2013**, *37*, 1075–1082. [CrossRef]
32. Pearlman, J.; Davis, M. Ecology of Fear: Los Angeles and the Imagination of Disaster. *Environ. Hist.* **1999**, *4*, 441–442. [CrossRef]
33. Guirk, P.; Dowling, R.; Maalsen, S.; Baker, T. Urban governance innovation and COVID-19. *Geogr. Res.* **2020**, *59*, 188–195. [CrossRef]
34. Acuto, M. COVID-19: Lessons for an Urban(izing) World. *One Earth* **2020**, *2*, 317–319. [CrossRef]
35. Acuto, M.; Dickey, A.; Butcher, S.; Washbourne, C.L. Mobilising urban knowledge in an infodemic: Urban observatories, sustainable development and the COVID-19 crisis. *World Dev.* **2021**, *140*, 105295. [CrossRef] [PubMed]
36. Nwajiaku-Dahou, K.; El Taraboulsi-McCarthy, S.; Menocal, A.R. *Fragility: Time for a Rethink*; ODI: London, UK, 2020; Available online: https://odi.org/en/insights/fragility-time-for-a-rethink/ (accessed on 13 June 2021).
37. Roitman, J. *Anti-Crisis*; Duke University Press: Durham, NC, USA, 2013.
38. Alraouf, A.A. The new normal or the forgotten normal: Contesting COVID-19 impact on contemporary architecture and urbanism. *Archnet-IJAR Int. J. Arch. Res.* **2021**, *15*, 167–188. [CrossRef]

Article

The Climate Just City

Mikael Granberg [1,2,3,*] and Leigh Glover [1]

1. The Centre for Societal Risk Research and Political Science, Karlstad University, 651 88 Karlstad, Sweden; leigh.glover@kau.se
2. The Centre for Natural Hazards and Disaster Science, Uppsala University, 752 36 Uppsala, Sweden
3. The Centre for Urban Research, RMIT University, Melbourne, VIC 3000, Australia
* Correspondence: mikael.granberg@kau.se

Abstract: Cities are increasingly impacted by climate change, driving the need for adaptation and sustainable development. Local and global economic and socio-cultural influence are also driving city redevelopment. This, fundamentally political, development highlights issues of who pays and who gains, who decides and how, and who/what is to be valued. Climate change adaptation has primarily been informed by science, but the adaptation discourse has widened to include the social sciences, subjecting adaptation practices to political analysis and critique. In this article, we critically discuss the just city concept in a climate adaptation context. We develop the just city concept by describing and discussing key theoretical themes in a politically and justice-oriented analysis of climate change adaptation in cities. We illustrate our arguments by looking at recent case studies of climate change adaptation in three very different city contexts: Port Vila, Baltimore City, and Karlstad. We conclude that the social context with its power asymmetries must be given a central position in understanding the distribution of climate risks and vulnerabilities when studying climate change adaptation in cities from a climate justice perspective.

Keywords: just city; climate just city; 'the right to the city'; climate change adaptation; power; equity; urban planning

Citation: Granberg, M.; Glover, L. The Climate Just City. *Sustainability* **2021**, *13*, 1201. https://doi.org/10.3390/su13031201

Academic Editor: Ingemar Elander
Received: 1 January 2021
Accepted: 21 January 2021
Published: 24 January 2021

Publisher's Note: MDPI stays neutral with regard to jurisdictional claims in published maps and institutional affiliations.

Copyright: © 2021 by the authors. Licensee MDPI, Basel, Switzerland. This article is an open access article distributed under the terms and conditions of the Creative Commons Attribution (CC BY) license (https://creativecommons.org/licenses/by/4.0/).

1. Introduction

Cities around the world are increasingly impacted by climate change and in total are forecast to experience massive social and economic losses under current trends [1–4]; by any reasonable understanding of the word, this constitutes a crisis [5]. This climate crisis threatens urban settlements [1] at the same time as the number of urban dwellers, in part due to ongoing urbanisation in higher-risk cities in low- and middle-income nations, exposed and vulnerable to climate-related events increases [6]. Future predictions state that climate change will increase the occurrence and severity of weather and climate-related hazards to urban settlements [1,2]. Climate change can also worsen existing socio-economic problems and the risks of other natural hazards.

Hence, issues connected to, or driven by, climate change are receiving increasing attention in cities around the world [7,8], and there is also unprecedented social mobilisation around these issues that is increasingly framed around the recognition that climate change is fundamentally a question of justice (cf., [9–11]). Justice, from this perspective, includes issues such as responsibilities for action, the varied vulnerabilities to the impacts of climate change, and how this is connected to structural injustices and power asymmetries that influences cities', organisations', and individual's abilities to mitigate risks and adapt to unavoidable impacts (cf., [6]).

Cities, therefore, are critical sites for the implementing measures to fulfill goals both for sustainability [12] and for economic growth [13]. At the same time, cities are dynamic " ... with interacting and interdependent social-economic, ecological-biophysical, and technological infrastructure components ... " ([14], p. 99). From the perspective of climate change, cities are perceived both as central problems and as important drivers of sustainable development with regard to climate change and other issues [14–18]. There are many demands, expectations, opportunities, and structures in the quest for the sustainable city [19]. Efforts to reorient cities under the rubric of sustainability are, at least in part, often framed by economic rationalities [20,21]. In this sense, climate change adaptation in cities takes place in what Suzanne Mettler has labelled a 'policyscape' consisting of layers of earlier political ideas, policies, and regulations that has become institutionalised with tangible " ... consequences for governing operations, the policy agenda, and political behavior" ([22], p. 369). A feature of this policyscape is path-dependency created by previous policy-making that facilitates some pathways and delimits other potential pathways of contemporary and future policy making [23,24]. Such path-dependency delimiting public governance and institutional activity is, of course, highly problematic, particularly when in a self-reinforcing condition, creating a 'lock-in' of extant governance. Such a lock-in in traditional governance models leads to " ... plans, policies and solutions that prioritize short-term goals over long-term resilience goals" ([14], p. 106). In order to reach transformative and climate governance models and frameworks facilitating climate justice in our cities, this path-dependency needs to be up-ended [14,25].

These circumstances highlight the importance of the issues of equity and fairness, decision making, participation and representation, power and influence, and governance [25–27]. In this article, we employ a critically constructive approach to the just city concept as developed by Susan Fainstein [28] and others, as we increase the complexity of the approach by adding the socio-environmental dynamics of climate change adaptation and focus on the 'climate just city'. Two cross-cutting topics central to a discussion of the climate just city with the inclusion of climate change adaptation emerges strongly from the discussion above (cf., [29–32]): (1) issues of distributional (or allocative) justice, and (2) issues of participatory (procedural) justice involving the processes, institutions and implications of citizen and stakeholder engagement [10,33,34].

These two aspects involve what Henri Lefebvre, David Harvey and others have termed 'the right to the city' [35–38], giving rise to political inquiry, such as: for whom is the city built and developed? What rights to the city as a whole do its inhabitants have? Who is given 'voice' in central political processes and whose experiences and knowledge is counted in the future development of urban society? Central for Lefebvre was the eradication of poverty and of unjust inequality [37,38]. Our highlighting of Lefebvre and Harvey also indicates that our outlook on the climate just city is more expansive than that of Fainstein's view of the just city.

This paper critically discusses the just city concept in a climate adaptation context and develops the just city concept by describing and discussing key theoretical themes in a politically and justice-oriented analysis of climate change adaptation in cities. We illustrate our arguments by looking at recent case studies of climate change adaptation in three city contexts: Port Vila (Vanuatu) [39], Baltimore City (USA) [40] and Karlstad (Sweden) [41]. Our cases are situated in different national and local contexts and highlight different aspects of the climate change adaptation challenge from the perspective of justice. Port Vila draws attention to the importance of the inclusion of indigenous populations and their experience-based knowledge in adaptation policy and its implementation; Baltimore City highlights how historical social class and race divisions impact contemporary adaptation policy and implementation. Karlstad brings to light how economic pressures and policy priorities frame adaptation policy and implementation. Before we move on to these case studies, we discuss cities and climate change adaptation, growth-oriented development policies and expand our theoretical perspective on just cities and that of the climate just city.

2. Cities and Climate Change Adaptation

Adaptation to climate change is essential for preparing for its impacts (as proactive or precautionary measures) and for responding to impacts that have occurred or are occurring (as reactive measures) (see, e.g., [26]). Measures for adaptation vary greatly, covering behaviour change and education, engineering and infrastructure, and institutional responses in law, policy and regulation, and, of course, planning (see, e.g., [1,7]). Furthermore, the goals for adaptation are wide-ranging, and include reducing exposure to hazards, hazard reduction, and vulnerability reduction. Key areas of urban adaptation related to the public sector include the energy, health, transport, water supply and sanitation portfolios [1,42]. As recorded by the UNFCCC, there has been greatly increased national activity globally reported in National Adaptation Programmes of Action and National Adaptation Plans; adaptation research dealing with cities has also similarly increased (see, [1]) together with associated scholarship [4,43–45].

Critically, the social and economic impacts of climate change are fundamentally unfair and unjust in their effects and consequences. A combination of societal and economic factors connects adaptation responses to issues of social (and environmental) justice. Climate change impacts produce social and environmental impacts of great scale, with an extensive proportion of the global population at risk of significant losses to essentially all components of social life; those at greatest risk are of lower socio-economic status, concentrated in lower- and middle-income nations and living in cities (around one-third of global population) [1,46]. Vulnerability to the risks of climate change is skewed against those with the fewest material resources and economic opportunities, greatest exposure to natural hazards (often tied to geographical location), and where economic/institutional circumstances inhibit adaptation capacities (see, e.g., [1,6,47]).

As a broad, dispersed, and ambiguous theme, estimating the scale of adaptation requirements/needs is difficult for any jurisdiction. A prominent problem is that although these costs are large and growing, they have typically been under-estimated. An associated problem is that investments in adaptation are clearly far below the level required to avoid great environmental, economic, and social losses. UNEP [48] estimates the costs of adaptation in developing nations alone by the year 2030 to be in the range of USD 140–300 billion/annum. UNEP's The Adaptation Gap Report [48] states that adaptation costs in developing nations to be probably two-to-three times than current global estimates by 2030. Óh Aiseadha, Quinn [49] found global adaptation policy expenditure for 2011–2018 to be USD 190 billion and the Climate Policy Initiative [50] estimated 2017/2018 expenditure at USD 30 billion. A report by the multilateral development banks of their collective climate finance in 2019 identified USD 15 billion adaptation expenditure [51].

As the major site of greenhouse gas emissions (GHG), cities are central to the goal of emissions mitigation. Emissions reduction globally, and the rate of reduction, will influence the scale and scope of climate change impacts and, in turn, that of the potential adaptation tasks facing cities. Accordingly, cities have a pivotal role to play in handling the climate change crisis, both in their efforts to limit negative and severe impacts in the future and also in combatting the already tangible impacts [25]. It has been stated that "...cities are the key human component in anthropocentric climate change" ([52], p. ix).

Adaptation is challenging in many dimensions; its definition is contested, it engages broad swathes of society and with such an extensive set of climate change impacts and associated values involved, adaptation responses canvass a potentially enormous field of activity with many economic, environmental, and social implications [26,53]. Significantly, adaptation is a highly heterogeneous phenomenon; for example, engaging individuals, households, and enterprises responding to local risks, major corporations adapting to changing production inputs, local governments planning for risks of coastal inundation and national governments formulating national policies, and a myriad of other issues, stakeholders, and responses. Within the diversity of adaptation awareness, approach, level of support, measures taken, stakeholder involvement and other variables is the lesson that adaptation is a locally situated phenomenon; it is a specific action taken in a defined location by stakeholders seeking to protect values in that site or related to that site (in the case of protecting ecosystem services). However, adaptation also poses 'cross-scale challenges' entailing multi-level governance [54–56]. As Moser expressed it: ([57], p. 31)

Those involved in organising, shaping, steering and implementing these efforts will have to navigate and manage a system made up of multiple actors with a variety of interests, capacities, and challenges often spanning several sectors. Moreover, many (if not most) locally planned adaptation decisions and actions require assistance from, or at least coordination with, higher levels of government—thus bringing additional actors to the table.

Accordingly, urban adaptation engages responses by civil society, private enterprise and governments, but with governments being the most prominent [26]. Public institutions play the major role in the urban climate change adaptation response through many functions, including disaster preparations and post-disaster recovery, environmental protection, land use planning, public goods infrastructure and services investment and provision, public education of disaster risk and preparation, public hazard protection and emergency service provision, relevant law and regulation formulation and enforcement, research and information gathering, social welfare provision, and urban development and building regulation. Government policy, programs, and planning have been recognised as essential in the adaptation response (in part, because of governments' unique capacities to deal with market externalities, market failures, protection of common pool resources and related problems), (see, e.g., [26]). Not only is adaptation a policy and planning challenge entailing technological and environmental adjustments to cities (such as through communication, housing, infrastructure and technology investments), but it is also a highly social and political one (including aspects such as equity, inclusion and exclusion, participation/representation, poverty and race) [40,58,59].

Climate change adaptation is one of several major challenges facing cities and it intersects, integrates, and overlaps with the problems of disaster hazards, unsustainable resource and ecosystem service use, and of underdeveloped nations and locations. United Nations policy responses at the international level provide a ready guide to this constellation of issues facing cities: the New Urban Agenda [60], Paris Agreement on Climate Change [61], Sendai Framework for Disaster Risk Reduction [62], Sustainable Development Goals [63] and the World Humanitarian Summit: Agenda for Humanity [64]. At the local, regional, national, and international scales of public policy and NGO responses, there are efforts to formulate concepts and policies that integrate elements of these separate agendas, such as the low-carbon city, the resilient city, and the sustainable city. It should not be assumed, however, that at the international—or any other—level, these policy and planning initiatives on related issues are aligned; rather, there will be a range of compatibility from the matching to the enabling to the constraining to the counteracting [18].

It is a truism in sustainability policy and planning that the greater the social transformation sought, the higher the barriers to change [45,65]. Adaptation barriers include the cultural and social, educational, financial and resourcing, informational, institutional, technological and the political, (noting that different stakeholders vary in their perception and understanding of barriers). Underpinning many of these barriers and constituting a 'mega barrier' is the manifestation of global capitalism (see several of the contributions in, [25]). For some scholars, the state's relationship with capitalist economics is a major barrier to formulating and implementing transformative adaptation measures (see, e.g., [32,66]). Critical for analysing adaptation is appreciating its relationship to capitalism, including such aspects as the role of economic markets, the expressions of economic interests, the arrangement of economics and political institutions, the activities of economic stakeholders, the economic valuations of social and environmental values, transfers from the public to private realms, market failures, access to common pool resources and the distribution of the costs of externalities. Government is intertwined with capitalism in the modern state, and while this relationship assumes many forms, it is indisputable that national governments assume responsibility for national economies, so that all national economies are engaged with global markets to varying degrees. National, state, and local governments have accepted, therefore, specific (albeit, highly varied) responsibilities regarding the performance of the economy under their jurisdiction; for elected representative governments, buoyant economic conditions are often tied to electoral success, and vice versa.

Taken together, climate change adaptation in a city context offers both considerable challenges and perceived opportunities, which clearly highlights the need to include political, social, and justice aspects when cities respond to the hazards and other impacts related to a changing climate. Until recent times, however, climate change adaptation has been informed primarily by science and technology, and it is only more recently that the academic adaptation discourse has widened to include social and cultural perspectives, as well as political analyses and critiques [26,59,67].

This emerging development in adaptation research approaches entails an increasing interest in issues such as the allocation and distribution of the costs and benefits of economic activity within society, the processes of decision-making and who is included/represented in decision-making institutions and who is excluded, and of the processes and institutions involved in decision making and the principles and directives that guide this activity [25,26,59,68–70]. Or, in a more bluntly critical formulation, those carrying the most profound risks and bearing the brunt of the negative impacts to date have their 'skin in the game'; some scholars have concluded that the public sector officials, technical advisors and elected representatives formulating adaptation responses are not likely to share the risks and potential pain of the most vulnerable and negatively affected by climate change [10,71]. In line with this reasoning, it has been stated that previous and contemporary urban climate change adaptation measures tend to " ... privilege existing engines of urban economic growth ... " ([72], p. 17) and, as a result, seek to protect those areas, assets and services, groups and individuals that are 'valued' from an economic growth perspective (see, e.g., [32,44]). These ideas were further developed in Anguelovski and Pellow [72].

3. Attractive Cities and City Growth

At the same time as cities are affected by climate change, they are also the primary engines of economic growth [73]. Changes in urban form, function and population are responses to economic growth, the most obvious expression of which is urbanisation in developing nations and the transformation from agrarian to industrial-service economies. In a form of feedback, urban growth can also foster economic growth and productivity increases through agglomeration, economies of scale and positive spill-over effects. Society in general and cities in particular are re-developing in the wake of changing patterns of industrialisation, urban forms and functions as driven by local and global economic and socio-cultural influences [41,52,74]. Globalization continues to draw urban settlements into the global economy, simultaneously providing new avenues for resource exploitation and market expansions to foster economic growth and also expanding the reach of market forces into social and economic life and exposing them to market competition.

Although global market forces are identified as producing net economic growth, a portion of economic gain is made at the expense of economic losses, such is the character of these markets [75]. Cities have become the locus, therefore, for the costs of globalization and of economic growth that include environmental and natural resource losses, increased economic marginalization and socio-economic 'precariousness', widening socio-economic inequality, and social unrest and conflict. Economic development, with its focus on competitiveness, can produce both industrial development in some cities and de-industrialisation in others, but in both scenarios, the city functions as a 'growth machine' (cf., [76]). In the case of developed nation cities in economic decline (or parts thereof), globalisation has initiated processes of city redevelopment focusing on the transformation of derelict urban areas into growth-promoting assets [13]. Poverty reduction and improvements in quality of life are the expected outcomes of economic growth in developing nation cities, but the association between growth rates and these outcomes varies greatly between cases; most conventional models of economic development link, to varying extents, domestic economic growth to international trade and globalisation.

Connected to this, local action on climate change, framed by perspectives of competitiveness and growth, has become a part of city marketing and branding with the aim of attracting investments and new inhabitants. As a result, a climate impact solutions-oriented approach that " ... offers a strategic opportunity that presents new pathways for investment, promotion, and regulation" ([73], p. 994) has evolved. Urban governance must contend, therefore, with the responsibilities it has assumed regarding managing economic markets and those it has assumed for managing, monitoring, planning, and undertaking adaptations to climate change.

4. Just Cities

In many respects, the just city concept is a trope used to simplify and analogise a highly contestable and uncertain idea that can be either positive (as a measurable condition or attribute) or normative (as an aspiration or specific goal). While usually viewed as concerning social justice in an urban context, the just city problematises two realms of high debate into one reflecting these uncertainties, namely the identity of the city (albeit frequently neglected in the just city discourse), the identity of justice and the positive and normative expression of both in the 'just city'. As a rhetorical or literary device, the just city frees debate and policymaking from the miasma of open-ended definitional debate, but for the potential price of fundamental uncertainty or false presumptions of shared outlooks.

Urban settlements as a generalisable form are elusive, with differentiations between the urban and non-urban being contested (see, [77]). Accordingly, there are differing views over the identity of just cities based in differences in the understanding of cities. Soja [78] argued for 'spatial justice', with its concern with how urban spaces are locations of social production. Chatterton [79] made the more expansive point that the city is a component of justice, not just its setting. It may be, therefore, that the expressions of injustice and opportunities for justice are more particular within specific urban areas than they are in the generic differences between the range of urban and non-urban forms.

Although the just city is a relatively new term, it represents ancient aspirations, debates and ideals over the character and meaning of justice. Notions of the just city can be found in the roots of Western philosophy and are certainly prominent in other philosophical traditions (see, e.g., [80]), with the roots of contemporary just city debates arising with the urbanisation of the Industrial Revolution (see, e.g., [81]). Although the urban identity component of just cities is a minor aspect of contemporary just city discourse, much of what has occurred under the just city moniker arose from the urban studies movement of the 1960s and 1970s in urban geography, urban politics, and urban sociology.

Fainstein [82] argued that, in the 1990s, the cause of justice was taken up more explicitly by scholars through three main approaches: (1) communicative rationality, (2) recognition of diversity and (3) the just city/spatial justice. Just city concepts were coincident with the broader outlook in left-wing politics, where the pursuit of justice has come to feature issues of representation and democracy alongside the longstanding concerns of income and material equity. Urban justice, wrote Fainstein [28], has the hallmarks of material equality, diversity and democracy. Promotion of the 'right to the city' by Lefebvre [83] and Harvey [35] entailed a more radical and critical cause than what Fainstein suggests:

The right to the city is, therefore, far more than a right of individual access to the resources that the city embodies—it is a right to change the city more according to our heart's desire. It is, moreover, a collective rather than an individual right, since changing the city inevitably depends upon the exercise of a collective power over the processes of urbanization. ([35], p. 939)

Harvey [36] expressed concern that Fainstein's concept of the just city was unquestioning of capitalism and avoiding aspects of outright conflict and struggle. In addition, and central for this article, Fainstein's just city concept fails to take socio-environmental dynamics into account, and as a development of the just city concept, we will delve more deeply into the implications of thinking of the just city in terms of the climate just city below.

5. Climate Just Cities

Cities around the world are increasingly impacted by climate change and it is increasingly evident that top-down action from the international down to the local scale will not suffice [5,84]. Cities have to play an important role in mitigating future GHG emissions and in adapting to contemporary and future impacts of climate change [1,3,4,45,52,54,85–87]. Policy and planning responses to climate change occur within the specific socio-economic and geographical milieu of each city. Cities, therefore, mirror society with its asymmetries in resource and power allocation, unequal distribution of burdens and suffering and skewed patterns of political participation and influence [40,88,89].

In this multi-level governance setting, the policy environment has become both increasingly complex, more differentiated at the city level, and also more responsive to international influences. Climate change objectives in cities are increasingly connected to broad policy objectives and goals, such as the UN's Sustainable Development Goals (SDGs), especially SDG 13 on climate action, but also to SDG 11 on making cities inclusive, safe, resilient, and sustainable [90]. Hence, addressing climate change in cities entails a social mix that gives rise to vulnerabilities, as it is saturated with substantial equity issues [26]. This means that climate change adaptation in cities is a highly political endeavour. Central questions from a climate just city perspective include: whose interests are being served and whose interests are being neglected? What are the social costs of adaptation and who carries them? Who can influence adaptation policy and who is excluded from processes of influence?

Although climate change adaptation, in a sense, is a universal task (cf., [91]), it is "... interpreted, localised, and modified in different settings ... locally translated through practice" ([89], p. 6). This means that climate change adaptation can be given different social and political meanings that, in turn, have different social and political impacts in different contexts [4,92]. In environmental policy in general, cities, through local government, have a central role [86]. Local government has the closest proximity to the citizens and is also the level where climate change impacts have their most concrete manifestations (and cities are also an important source of GHG emission mitigation measures) [14,84,93–95].

Studies show that climate change adaptation is often enacted in a defensive manner, focusing on what needs to be defended or preserved rather than "... what can be reformed or gained" ([43], p. 3). Consequently, climate change adaptive measures very seldom deliver on their transformative potential (cf., [45]). In line with this argument, it has been stated that, depending of the pathway adopted, climate change adaptation can "... maintain a status quo and uphold existing political circumstances, whilst transformation entails political change. Hence, adaptation is enmeshed in both contemporary politics and those of the future" ([26], p. 14).

Bulkeley and Edwards [96] proposed that climate justice in the city conventionally is presented as having two axes, one of distributions–procedures and the other of rights–responsibilities, to which they add recognition as an underlying aspect of justice brought forward by the requirements of environmental justice. Schlosberg [97] argued for global climate justice and offered a trivalent view of the need for recognition, distribution and participation. This trilogy of justice components covers the majority of interests expressed in the 'just city' but may not be complete. Two contemporary environmental discourses can be used to complement the just city conception. Firstly, environmental justice added to notions of urban justice by explicitly viewing the environment as a source of 'goods and bads' that is mediated by cultural, geographical, and historical factors and needs that are distributed through political processes. Secondly, ecological justice also expanded the realm of justice by recognising the interests of environment (i.e., recognising an intrinsic valuation of natural phenomena); given the ecological footprint of cities, ecological justice widens the realm of justice in several dimensions.

More commonly, climate justice is taken to deal with distributive justice and procedural justice, the former concerned with income and resources/assets and the latter with involvement in decision making. Distributive justice comes into play when income, resources/assets, ecosystem services and the like interact with vulnerability to climate change impacts (cf., [89]). Adaptation measures can serve to secure the current distribution/allocations/ownership, resource and ecosystem service use and/or access under climate change or actively re-distribute these; such outcomes can be in anticipation of climate change or reactions to changes underway. Critically, adaptation concerns the distribution of the costs of adaptation (and avoided costs) and benefits of adaptation. Key themes in scholarship include the cultural, geographic, and socio-economic specificity of vulnerability and adaptation, equity, and fairness. Procedural justice concerns the role and powers of stakeholders (in civil society, corporations and government) in decision making in adaptation policy, planning, and practices.

Accordingly, the concept of climate justice overlaps with the just city, but there are some important differences. To some extent, these are rival discourses sharing considerable common thematic territory, and despite having a common root in the Western conceptual tradition of justice, any convergence in current interests is the result of having reached a similar destination through quite different scholastic routes. In one sense, climate justice can be an additional supplemental element to the just city; there does not appear to be any aspect of climate justice that might be considered to detract from, or otherwise weaken, the just city cause. Climate justice does have, however, a number of elements that distinguish it from the just city concept, goals, and discourse.

From a historical perspective, climate justice is the more recent conceptual arrival in scholarship. As is often the case, however, preceding this recent label is a considerable body of earlier work describing the relationship between society and climate, including climate change and its socio-economic and cultural impacts [10,11,96,98]. Temporal issues distinguish between climate justice for adaptation and the just city, with the former being particularly concerned with future change (of the climate and climate-related systems); just cities have concentrated on historical antecedents and contemporary conditions, whilst climate justice necessarily also entertains speculations and forecasts of future change. The scope of interest can also be a distinguishing feature, with the just cities discourse tending to concentrate on socio-economic activities within the confines of the urban unit, whereas just climate scholarship often envisages the city as functioning within a broader economic and cultural system well beyond its nominal or formal boundary (following the insights of the ecological footprint concept).

As a discourse, climate justice has moved fairly quickly to overlap and exchange influences with those of disaster risk studies and sustainable development. This is due, in large part, to much of the work on climate justice having an interest in environmental politics, notably the scholarship and activism in environmental justice. A component of this merger is exemplified in socio-economic assessments of natural disasters related to climate (and weather) in cities, despite the complication of uncertainty over the relationship of single events to climatic change; such as into the political components of the vulnerabilities, risks and outcomes of the effects of Hurricane Katrina on New Orleans [99].

Climate justice (and climate politics) arises from the particular institutional circumstances of the international response to climate change that centres on the activities and outputs of the UN as played out in the international and sub-international jurisdictional realms. This provides the contrast between the single-issue origins of climate justice and the more diffuse origins of the just city discourse and scholarship. Therefore, climate justice concerns a particular set of policy actors, institutions, interests, markets, policies, and relationships. A feature of climate politics is the extent to which these international features exert an influence on the climate policy response at all levels of governance, including that of cities. Two other defining aspects of climate justice in adaptation spring from this factor. Firstly, for many poorer nations, cities and communities, support from international bodies and other external sources will be critical for undertaking effective urban adaptation measures (see, e.g., [1,46,48,61]). Secondly, climate justice often seeks linkages between adaptation responses and measures to promote GHG emissions reductions, sometimes envisioned as transforming cities into low-carbon settlements; in other words, climate justice can be positioned as a component of socio-ecological transformation (see, e.g., [25,27,43,72]).

As such, our view of the climate just city is one with both positive and normative elements; it is, therefore, not only an aspirational goal or utopian ideal concept, but also focusing existing conditions, achievements, and victories in overcoming oppression and injustice.

6. Three Case Studies of City Climate Change Adaptation and Vulnerability

Below we will present three case studies that illustrate different aspects of climate change adaptation and climate (in-)justice in three cities as a way to fuel our discussion on the climate just city.

6.1. Port Vila

This case study reviews the factors and circumstances shaping the adaptation response and associated justice implications in the urban and peri-urban areas of Port Vila, the capital of Vanuatu, a Melanesian small island developing state (SIDS) in the western Pacific Ocean (the text under this heading is primarily based on [39]). Urban vulnerability has been largely neglected in SIDS scholarship and international development initiatives; however, recent scholarship has highlighted the scale and scope of climate change risks to urban settlements. Although not always identified with urban settlements, in a large number of cases, the more populous SIDS are relatively highly urbanized, with a high proportion of housing in densely settled and extensive informal settlements, and with high urban growth rates. Coastal locations, high risk of natural hazards, a narrow economic foundation, weak urban infrastructure investment, and low average per capita incomes typify the factors contributing to the vulnerability of SIDS urban settlements to climate change impacts (especially those associated with extreme weather events), conditions also applying to Port Vila [100]. Climate change (and the consequences of natural hazards, notably tropical cyclones) appears to be further promoting urbanization and rural-to-urban migration to the city. In 2015, Port Vila was identified as one of the most natural disaster-exposed out of a global sample of 1300 cities [47], notably in the face of the scale of economic losses caused by Cyclone Pam in 2015 relative to its GDP.

Port Vila's vulnerable urban populations highlight a set of causal factors that are essential to an effective program of adaptation, namely those of governance, institutions, and participation/representation in adaptation planning, policy formulation, and financing [101]. One indication of the problem of governance in the city's capacity to adapt is that although the official 2009 city population was around 45,000, another 15,000 or so citizens occupied the city's informal settlement and peri-urban areas; similarly, the official population growth rates significantly underestimated the rates for the settlement as a whole [102]. Differences between official recognition and non-recognition are representative of the formal and informal divisions within the city; the former is a part of formal government and the latter has both governmental and extra-governmental functions and structures, including NGOs, 'quasi-customary' institutions, and churches. As described above, adaptation policy and practice are a response to climatic hazards, the exposure risk of entities, services and values to these hazards, and the deployment of the ability to respond (covering culture, governance, finance, technology and many other variables). There are implications for adaptation responses in Port Vila, therefore, as a consequence of this administrative, political, and social bifurcation.

Port Vila is vulnerable to climate change and natural hazards partly because of the condition and capabilities of its institutions for governing and managing urban planning, urban services management and delivery, and disaster risk reduction. Growth in the city is concentrated in the peri-urban areas, a factor that exacerbates the city's vulnerability/adaptation dichotomy, as these peri-urban areas lack formal urban governance and are disconnected from local government services and infrastructure. Informal settlements, therefore, pose particular challenges for adaptation governance for public agencies, in which land tenure can be critical (and where, in practice, tenure assumes numerous and complex forms). Mitchell and McEvoy ([103], p. vii) found that adaptative capacity is limited by 'insecure tenure' as such communities are " ... disconnected from formal governance processes, lacking the knowledge and information for informed decision-making, and having restricted access to finance for implementing resilience-enhancing actions." An immediate implication for justice is that occupants of informal settlements face the twin disadvantages of increased risk exposure and high vulnerability combined with lower adaptive capacity and less state support in the form of engagement in capacity-building, infrastructure provision, planning measures, and resources/investments compared with formal settlement areas.

In Port Vila's case, the problem of informal settlements and unchecked peri-urban settlement reflects the legacy of a post-colonial condition, as under colonial rule, the efforts to build, own, and control the city actively worked against customary law and other customs within the city. Historical exposure to climatic variations and extremes has built high levels of community-based coping capacity, based on traditional knowledge and technologies, amongst some the city's inhabitants. Trundle [39] argued the need for understanding the underlying non-climate related social system such as governance and agency issues and its history and how this connects to contemporary climate action. This historical context has a direct impact on current conditions for governance, infrastructure conditions and urban planning and the understanding of this context is " ... critical if climate-related interventions are to be effective, relevant, equitable and sustainable" ([39], p. 37).

Due to the colonial heritage, however, these resources exist separated from the post-colonial urban institutions that form the urban governance system. This has also led to the peri-urban governance being " ... disjointed from Port Vila's established urban core" ([39], p. 37). Therefore, to a great extent, the vulnerabilities of peri-urban Port Vila is due to an adaptation deficit.

6.2. Baltimore City

This case study examines urban government climate action, risk reduction, and preparedness in Baltimore City, in the United States (noting that Baltimore City is within the larger Baltimore metropolitan area) (the text under this heading is primarily based on [40]). This case study aims to increase the understanding of efforts to increase climate change resilience in a historically and contemporarily racially segregated city, a segregation that also over the years has had physical expressions with environmental privileges allocated to mostly white residents, such as planting of trees only in white neighbourhoods and limited access to green spaces in segregated areas [33]. The city's population is around 620,000, with 64 percent African American, 30 percent white, 4 percent Hispanic or Latino, and with 24 per cent of its population living below the federal poverty line [104]. Baltimore City is situated in the state of Maryland on the Chesapeake Bay, which is the largest and most diverse estuary in the United States [105]; it is the largest city in the state of Maryland.

Maryland has over 4000 miles of shoreline and the state is particularly susceptible to erosion, flooding, storms and, increasingly, sea level rise. Baltimore's location makes it vulnerable to a range of natural hazards, including coastal storms, extreme heat, flooding and high winds; studies have forecast increasing vulnerability to extreme events, notably flood hazard [106]. The city is also vulnerable to extreme heat, low air quality related to its urban heat island effect, and food insecurity [104,107,108]. Over the last decade, Baltimore has endured many severe weather events ranging from heavy precipitation, tidal floods, snow and ice storms, coastal storms, heat waves and even experienced a tornado in its Inner Harbour. Impacts from these events have impacted the city's residents, businesses infrastructure, and its natural systems.

Baltimore is a port city sitting on a waterfront at the intersection of four major watersheds. The city has experienced population decline since the 1950s and as a result, has invested heavily in revitalizing its downtown precinct [107]. A major feature of this investment program, and a high priority on the city's policy agenda over recent years, has been waterfront development and redevelopment. Although there are great commercial opportunities in the waterfront location, there are also significant challenges connected to the location. This redevelopment has resulted in a downtown concentration of workplaces but with a relatively modest increase in residential housing.

A socio-economic characteristic of the city is the significant pockets of extreme poverty and social vulnerability in close proximity to its downtown [107]. There is a racial signature to these impoverished districts—that of a concentration of minorities—that has persisted over time, as the city has history of deliberate segregation and is one of the most segregated cities in the United States. It also has an overall higher poverty rate than most US cities and its poorer areas are mainly African American. This is clearly observable in the environmental, economic, and social challenges the city faces today. As a result, the city's " ... unique combination of shocks and stresses cuts across social, economic and environmental factors" ([40], p. 128).

The participatory process in the city in climate change adaptation processes has been quite ambitious. However, there has been a strong focus on plans and planning, such as the Sustainability Plan, the Climate Action Plan, the Disaster Preparedness and Planning Project, a combined all hazards mitigation plan and climate change adaptation plan. There has been a lack, however, of movement beyond planning to action aiming at protecting the most vulnerable communities and properties from harm [107]. Facing such implementation failures by the public sector, there are opportunities for other actors to substitute for the state and even move beyond government action, thereby evoking the question of whether or not there is sufficient civic capacity to fulfill this role. Due to historically determined limits on the capacities for social self-organisation, the community considers climate change adaptation to be primarily a government responsibility. For this reason, citizens' interest in participation seems elusive.

This dilemma indicates the need to acknowledge the influence of racism on city government planning that has shaped contemporary disenfranchisement and inequities across Baltimore's communities, leading to a lack of trust in government actors and agencies by minority populations [104]. Therefore, the city has recognised the need to utilise an 'equity lens' for all climate change action. Despite this, the efforts had shortcomings, with difficulties keeping participants engaged throughout the entire process, especially volunteers. The city could have been more efficient in supporting the volunteers and formulated more realistic targets. More direct efforts were needed to ensure that not only race but also ability level, age, economic status, and other relevant socio-economic factors were considered. It was clear that integrating equity and climate change into everyday decision-making was extremely difficult.

6.3. Karlstad

This case study examines urban climate change adaptation planning in the city of Karlstad, Sweden (the text under this heading is primarily based on, [41]). Karlstad is a small but expanding city on northern Europe's largest lake and river delta, with about 94,000 inhabitants; the city has relatively high climate change ambitions [109]. Ongoing population growth has led to densification and to intense property development in the city [110]. The city has experienced a number of large floods over the years. Karlstad has also been identified by the State Commission on Climate and Vulnerability and the European Union as one of the Swedish cities most vulnerable to climate change-related flood impacts. The city is also a 'Resilient City' designated by the United Nations Office for Disaster Risk Reduction (UNDRR), and one of the most active Swedish cities in waterfront redevelopment. As the city is located on a river delta and by a lake, there is an established historical awareness of flood risks, and this awareness also includes potential contemporary and future impacts. This can be illustrated by an old proverb that says that when the river and the lake meet at the central square in Karlstad, this will be the end of the city [111,112]. The city's key strategic goals are growth, attractiveness, and to become a 'good' green city [110].

In the Swedish context, the municipalities have responsibility for spatial planning within their own territory and this gives them an extensive mandate in city development and redevelopment [112]. Central to Karlstad´s approach to city redevelopment is the vision 'Karlstad 100,000'. Its aim is to make the city attractive to development investments by utilizing the city's lake and riverine location. The approach focuses on business actors, competitiveness and growth coalitions. The main component is the redevelopment of the inner harbor. The vision's strategy is to attract the professional classes and other well-educated individuals and households to take up residency and that this, in turn, will attract and enable investment from private businesses.

The city has experienced a number of floods over the years. In 2000/2001, heavy rain that gradually increased the lake water level resulted in flooding the city, an event that that lasted 6 months, covering large parts of the city and leading to extensive damage and disruption. The city has been active in its development of its climate adaptation and risk reduction capacities. The city employed a flood plains manager in 2007, launched a flood risk management program in 2010, works continuously with stake-holder collaboration and outreach, and has built several physical barriers to prevent future flooding in the city.

Hence, the issues of city growth and climate change adaptation have developed in parallel over the years, but the city's economic development agenda has gradually evolved to being the number one priority, and subsequently framing climate change risk analysis and adaptation measures. The case study shows a growing awareness among city officials of this double challenge for planning and city redevelopment. However, this inherent conflict is unresolved and can be observed " … in the light of historical experiences, contemporary flood risks and the use of waterfront lands as an important component in developing city competitiveness" ([112], p. 33). Despite the awareness of the double challenge, it is not explicitly problematized in the policy agenda and appropriate measures to mitigate flood risk have not been able to compete with the objectives to construct 'urban attractiveness' as a means to increase economic competitiveness on the city's policy agenda.

As a result, competitiveness is the dominant city policy goal that defines the 'policyscape', limiting the space for other issues not perceived as compatible with this overarching aim. Climate-related risks discussed in the city planning documents are designed to accommodate flood risks in ways that facilitate city competitiveness. Pre-emptive risk reduction measures, such as retreat from the waterfront as a defensive climate adaptation measure and investing in other city areas less prone to flood risk, are not on the policy agenda. At the same time, the waterfront redevelopment is increasing the exposure to flood risks, and this will most likely increase due to climate change. This framing of climate vulnerabilities by competitiveness is potentially dysfunctional, as it can limit city resilience and increase the vulnerability of the city in both the short term and in the longer run.

7. Discussion: Implications for the Climate Just City

The impacts of climate change on climate justice can be both direct and indirect (see similar discussions in, [26]). Direct impacts are concrete and tangible effects on the material and social realms, such as those on disenfranchised areas, groups and individuals; indirect (or secondary) impacts are spill-over effects from direct climate impacts connected to institutionalised patterns in society co-creating cascading social and economic outcomes. Adaptation measures can also have negative consequences (i.e., maladaptation). Direct and negative (actual and potential) impacts can increase the vulnerability of social and socio-ecological entities and processes. Indirect negative impacts can be more diffuse and produce effects temporally and spatially distant from the site of direct impact, affecting other locations, sectors, systems and other realms of society. Both types of impacts are observable across the three cases.

The case studies highlight the interaction between indigenous and experience-based knowledge and expert-driven urban and adaptation planning, how historical disadvantages reinforce vulnerabilities to a changing climate and excludes social groups from the benefits of climate change adaptation, and how policy path-dependency in terms of city growth and competition policies produces risk and vulnerabilities and drive the need for climate change adaptation measures in certain areas, while ignoring other needs in other areas.

The Port Vila case study highlights the value of including indigenous and local communities and their knowledge in formal urban planning processes and problematizes the impacts when the inclusion of this type of knowledge fails. Studies show that climate change adaptation is mainly based in scientific and technological knowledge and "... views traditional and indigenous knowledge as insufficient for resilience in the face of new types and levels of climate hazards" ([26], p. 33). The Port Vila study problematizes arbitrarily drawn urban boundaries in managing climate change adaptation, building climate resilience, and investigating vulnerabilities. It points toward the need to include indigenous/traditional knowledge and narratives/imaginaries, community adaptation capacities, non-institutional means of coping with natural hazards and alternative pathways in order to handled vulnerabilities in all segment of society [18]. This deficiency in 'top-down' urban planning can lead to climate change adaptation planning and measures unable to deal with future climate risks and vulnerability reduction (cf., [26]). It is important to include communities in ways that facilitate social learning and makes integration of non-expert knowledge possible as this "... reinforces the understanding of the historical and cultural origins of place and of the locally specific informal arrangements they maintain and sustain" ([113], p. 155).

The Baltimore case demonstrates the need to include vulnerabilities and its long historical roots when designing climate change adaptation measures aiming at building new, and reinforcing existing, capacities of the most vulnerable areas, groups and individuals in the city and also taking into account what happens when this fails. The city government must have a commitment to addressing profound historically developed social and economic challenges and connect them to climate-related risk in order to support the most vulnerable in the city.

The Baltimore case also highlights the need for climate just adaptation to focus on equity (giving people the resources they need in relation to what they have been denied historically), rather than equality (i.e., treating everyone equally). It also demonstrates how the city's history and social and economic context has created vulnerable areas and communities. These areas are socially vulnerable in general but are especially vulnerable to climate change. Effective and just climate adaptation must, therefore, address issues of accessibility, lack of resources, racial inequality and social exclusion. A truly equitable process facilitating development towards a climate just city is demanding and requires "... honesty, real talk and a true recognition of one's own privilege" ([40], p. 126). This is a challenge for city planners that have to connect present-day circumstances to historical injustices in ways that involves continuous learning and acknowledgement.

Focusing climate change adaptation efforts towards more vulnerable individuals, groups and city areas can increase general resilience and has the potential to empower stakeholders, build trust and promote community cohesiveness. In general, the cases show (and especially evident in the cases of Port Vila and Baltimore) that failure to integrate 'top-down' climate action with 'bottom-up' knowledge will not adequately address needs and vulnerabilities in ways that facilitate resilience-building or climate justice.

The cases of Port Vila and Baltimore City also clearly show that climate change adaptation with the potential to contribute to a climate just city needs to have a comprehensive understanding of how different forms of climate change adaptation should take place in a specific, historically developed, economic, environmental, and social context (cf., [113]). A just climate change adaptation must therefore include an understanding of how success and failures are shaped by place-based specificities that have emerged over time, creating structures of power, impacting actors' capacities and individuals' and group's vulnerabilities. Thus, the historical city context has clear implications for actors' ability to be proactive or react to climate change and to withstand or adapt to its impacts. Attention must be given, therefore, to a wide range of socio-economic and cultural diversities in order to reach climate just outcomes. This requires an inclusive approach sensitive to cultural diversity, history and social inequality. Clearly, just climate change adaptation is not a separate challenge but has to address both the physical impacts of climate change and social injustices, i.e., to address climate justice.

The case of Karlstad highlights the climate justice problem of implementing climate change adaptation within the constraints of dominant policy priorities shaping city development goals and processes. For local climate action in general, high-level support is critical as it influences the positioning of climate issues on the political agenda and resource allocation [113]. This case highlights how climate change is given a place on the local policy agenda but is framed by other prioritized political objectives and how this, in turn, limits more effective policy responses to climate change.

Additionally, there is an inherent tension between 'business-as-usual' city growth policies and competition goals focused on waterfront housing developments on the one hand, and the need to manage climate-related risks related to future inundation and reduce vulnerabilities on the other. This tension is resolved, however, in the favour of (neoliberal) economic growth, a priority that frames local climate change policy and action and outweighs the potential impacts of climate change risks. Sometimes climate action in cities framed by neo-liberal growth policies is labelled 'climate urbanism' and there are distinctive social justice concerns resulting from this condition [73]. Consequently, city action to facilitate growth and competitiveness has clear social implications through heightened vulnerability and increases in risk exposure, that in turn increase the future burden of adaptation.

The Karlstad case is a clear example of gentrification, as the project involves restoring a contaminated site and developing a former industrial harbor to an upscaled housing and entertainment district [33,114,115]. This goal is facilitated and guided through top-down planning. Within the rhetoric of the project is a rationale to address social justice, namely that the harbor development will increase city attractiveness and generate growth that will 'trickle down' to the most vulnerable in the city [116]. This is, however, a contentious view that has been problematized in earlier scholarship as " ... "trickle down" policies ... are means whereby the nonpoor majority benefit at the expense of the poor" ([117], p. 291). That being so, gentrification is in itself problematic from a just city perspective, and the lack of bottom-up involvement in the planning process mainly channeled through elite political and administrative actors furthers this perception. Furthermore, the process in the Karlstad case can, from a climate change perspective, be perceived as 'green' gentrification, as it includes comprehensive climate change adaptation measures. However, this is also problematic, as studies of other locations have shown that " ... many green interventions create enclaves of environmental privilege when low-income and minority residents are excluded from the neighborhoods ... " and that, hence, " ... many greening projects remain blind to social vulnerabilities" ([33], p. 1065). This points to the fact that even so called 'win-win' climate change adaptation approaches and measures often obscure how uneven costs and benefits impacts different groups differently and leaving vulnerable groups more vulnerable [34,118]. This highlights the importance of asking the question: adaptation for whom?

8. Conclusions

In these cases of climate change adaptation, policy and action is mediated through the dominant policy priorities shaping the wider city development interests and goals [4,26]. The framing and subsequent policy formulation of the public sector adaptation response enables and constrains the problem-solving capacity of cities and is of central importance for understanding how local climate action is interpreted, framed, formulated and prioritised and how this impacts the inhabitants of a city. How climate change adaptation is framed in a specific policy context shapes how it is addressed. For instance, if urban planning decision-making integrates broader social development outcomes, it can facilitate the inclusion of justice issues into climate change adaptation policy and action [119]. As addressing climate change will always be a contested public policy process in which adaptation responses must compete with other political priorities and policy goals, the resulting decisions by governments and public agencies will involve compromises and trade-offs, so that the 'who gets what, why, when, and how' questions involve political themes of power relations, interest conflicts and related factors becomes of great importance when studying climate justice.

In a way, our thoughts on how a climate just city can be approached overlaps with what Peter Marcuse called 'the subversive reading´ of Lefebvre's call for the right to the city [37]. For a climate just city to be more than a utopian aspiration, transformative change must be realized in ways that meet the needs of both social and climate justice for excluded groups and for society as a whole [27,45]. To do this, equity issues must be at the forefront of government climate change adaptation responses, involving increasing public provision to fight poverty, reducing urban planning's focus on economic short-term gains and on city growth in narrowly cast economic terms, and increasing the involvement and inclusion of vulnerable groups in the political processes of the city. Considering our contemporary society, this is a substantial challenge and, as the three cases indicate, there is a considerable distance to go before reaching a transformative approach that could facilitate a climate just city.

From the perspective of the climate just city, a broad stakeholder collaboration is central for enhancing transformative capacity. Tempering this goal are the real-world circumstances of governments' interest in economic management, acceptance of economic globalisation and neoliberal economic policies that feature close ties, collaborations, mutual interests, social networks and informal governance arrangements with private actors that are stakeholders in the corporate sector. These set-ups can reinforce existing power asymmetries and influence and change democratic and administrative practices as they open up for forms of network governance with limited transparency and opportunities for democratic accountability. For this reason, the utilisation of knowledge, who is included and excluded in the decision-making process, and the impact on political priorities have the potential to make or break achieving a climate just city.

As expounded in the introduction and in the case studies, climate justice entails inquiries into the construction of justice and often recognises non-material values, such as inclusion/exclusion, participation and recognition of minorities [26,29,30,68]. Climate change adaptation poses challenges that evoke moral issues, and it constitutes an ethical challenge (cf., [120]). Adaptation responses are an overlay on existing social differences, with the potential to alleviate, be neutral or exaggerate these differences [40]. Key aspects of social difference inimical to social justice and achievement of the just city include socio-economic inequalities, limitations to participation in decision making, variability in exposure to risks and priority-setting in adaptation policies [26]; all aspects observable in our cases. These aspects affect the distribution and intensity of vulnerabilities to climate change impacts and are expressions of injustice and unfairness. Class, gender, and race and other expressions of social differentiation are structural components in social injustices exacerbating vulnerabilities to the impacts of a changing climate. All these aspects are, of course, central to the analysis of climate just cities and climate change adaptation.

A central aspect of the focus on a climate just city is the critique of the apolitical perspectives in much adaptation scholarship [66,121]. This critique argues that apoliticism restricts political inquiry by taking adaptation as the 'natural' response to ecological problems without considering its social and political implications [122,123]. Ignoring the political in governmental climate change adaptation responses obscures the role of processes—such as capital accumulation, technology change and political contestation—essential in producing the lived environment. It has been argued that this orientation suits " . . . the institutional needs for controlling the procedures of social change, thereby circumventing issues of power hierarchies, vested interests and the like" ([26], p. 146). Thus, as our cases clearly illustrate, the social context, with its power asymmetries, must have a central position in understanding the distribution of climate risks and vulnerabilities [124] when studying climate change adaptation in cities from a climate justice perspective.

Author Contributions: Both authors have worked on the conceptualization, the analysis and writing. All authors have read and agreed to the published version of the manuscript.

Funding: This research was funded by the research fund of the Swedish Civil Contingency Agency, MSB/2016-6855.

Informed Consent Statement: Not applicable.

Data Availability Statement: Not applicable.

Acknowledgments: The authors would like to thank the three anonymous reviewers for their valuable comments. The authors are also grateful for the support received from the Centre for Societal Risk Research, Karlstad University, Sweden.

Conflicts of Interest: The authors declare no conflict of interest.

References

1. IPCC. *Climate Change 2014: Impacts, Adaptations and Vulnerability*; Cambridge University Press: Cambridge, UK, 2014.
2. IPCC. *Global Warming of 1.5 °C. Geneva: IPCC (Intergovernmental Panel of Climate Change)*; Cambridge University Press: Cambridge, UK, 2019.
3. Bulkeley, H.; Castán Broto, V.C. Government by experiment? Global cities and the governing of climate change. *Trans. Inst. Br. Geogr.* **2013**, *38*, 361–375. [CrossRef]
4. Moloney, S.; Fünfgeld, H.; Granberg, M. (Eds.) *Local Action on Climate Change: Opportunities and Constraints*; Routledge: London, UK, 2018.
5. Glover, L. *Postmodern Climate Change*; Routledge: New York, NY, USA, 2006.
6. IFRC. *World Disaster Report 2020: Come Heat of High Water*; International Federation of Red Cross and Red Crescent Societies: Geneva, Switzerland, 2020.
7. Olazabal, M.; Ruiz de Gopegui, M.; Tompkins, E.L.; Venner, K.; Smith, R. A cross-scale worldwide analysis of coastal adaptation planning. *Environ. Res. Lett.* **2019**, *14*, 124056. [CrossRef]
8. Castán Broto, V.; Bulkeley, H. A survey of urban climate change experiments in 100 cities. *Glob. Environ. Chang.* **2013**, *23*, 92–102. [CrossRef] [PubMed]
9. Newell, P.; Srivastava, S.; Naess, L.O.; Torres Contreras, G.A.; Price, R. *Towards Transformative Climate Justice: Key Challenges and Future Directions for Research Contract No.: DS Working Paper 540*; Institute of Development Studies: Brighton, UK, 2020.
10. Porter, L.; Rickards, L.; Verlie, B.; Bosomworth, K.; Moloney, S.; Lay, B.; Latham, B.; Anguelovski, I.; Pellow, D. Climate justice in a climate changed world. *Plan. Theory Pract.* **2020**, *21*, 293–321. [CrossRef]
11. Edwards, G.A.S. Climate justice. In *Environmental Justice: Key Issues*; Coolsaet, B., Ed.; Earthscan: London, UK, 2021; pp. 148–160.
12. Fenton, P.; Gustafsson, S. Moving from high-level words to local action—governance for urban sustainability in municipalities. *Curr. Opin. Environ. Sustain.* **2017**, *26–27*, 129–133. [CrossRef]
13. Galland, D.; Hansen, C.J. The roles of planning in waterfront redevelopment: From plan-led and market-driven styles to hybrid planning? *Plan. Pract. Res.* **2012**, *27*, 203–225. [CrossRef]
14. McPhearson, T. Transforming cities and science for climate change resiliencein the anthropocene. In *Transformative Climate Governance: A Capacities Perspective to Systematise, Evaluate and Guide Climate Action*; Hölscher, K., Frantzeskaki, N., Eds.; Palgrave Macmillan: London, UK, 2020; pp. 99–111.
15. McCormick, K.; Anderberg, S.; Coenen, L.; Neij, L. Advancing sustainable urban transformation. *J. Clean. Prod.* **2013**, *50*, 1–11. [CrossRef]
16. Parnell, S. Defining a global urban development agenda. *World Dev.* **2016**, *78*, 529–540. [CrossRef]
17. Krellenberg, K.; Bergsträßer, H.; Bykova, D.; Kress, N.; Tyndall, K. Urban Sustainability Strategies Guided by the SDGs—A Tale of Four Cities. *Sustainability* **2019**, *1*, 1116. [CrossRef]
18. Westman, L.; Castán Broto, V. Urban climate imaginaries and climate urbanism. In *Climate Urbanism: Towards a Critical Research Agenda*; Castán Broto, V., Robin, E., While, A., Eds.; Palgrave Macmillan: London, UK, 2020; pp. 83–95.
19. Cohen, S. *The Sustainable City*; Columbia University Press: New York, NY, USA, 2018.
20. Martin, C.J.; Evans, J.; Karvonen, A. Smart and sustainable? Five tensions in the visions and practices of the smart-sustainable city in Europe and North America. *Technol. Forecast. Soc. Chang.* **2018**, *133*, 269–278. [CrossRef]

21. Haughton, G. Environmental justice and the sustainable city. *J. Plan. Educ. Res.* **1999**, *18*, 233–243. [CrossRef]
22. Mettler, S. The policyscape and the challenges of contemporary politics to policy maintenance. *Perspect. Politics* **2016**, *14*, 369–390. [CrossRef]
23. Peters, B.G.; Pierre, J.; King, D.S. The politics of path dependency: Political conflict in historical institutionalism. *J. Politics* **2005**, *67*, 1275–1300. [CrossRef]
24. Patterson, J.J. Institutional dynamics of transformative climate urbanism: Remaking rules in messy contexts. In *Climate Urbanism: Towards a Critical Research Agenda*; Castán Broto, V., Robin, E., While, A., Eds.; Palgrave Macmillan: London, UK, 2020; pp. 97–115.
25. Castán Broto, V.; Robin, E.; While, A. (Eds.) *Climate Urbanism: Towards a Critical Research Agenda*; Palgrave Macmillan: London, UK, 2020.
26. Glover, L.; Granberg, M. *The Politics of Adapting to Climate Change*; Palgrave Macmillan: London, UK, 2020.
27. Hölscher, K.; Frantzeskaki, N. (Eds.) *Transformative Climate Governance: A Capacities Perspective to Systematise, Evaluate and Guide Climate Action*; Palgrave Macmillan: London, UK, 2020.
28. Fainstein, S.S. *The Just City*; Cornell University Press: New York, NY, USA, 2011.
29. Schlosberg, D. Theorising environmental justice: The expanding sphere of a discourse. *Environ. Politics* **2013**, *22*, 37–55. [CrossRef]
30. Coolsaet, B. (Ed.) *Environmental Justice: Key Issues*; Earthscan: London, UK, 2021.
31. Walker, G. *Environmental Justice: Concepts, Evidence and Politics*; Routledge: London, UK, 2012.
32. Sovacool, B.K.; Linnér, B.-O. *The Political Economy of Climate Change Adaptation*; Palgrave Macmillan: London, UK, 2016.
33. Anguelovski, I.; Brand, A.L.; Connolly, J.J.; Corbera, E.; Kotsila, P.; Steil, J.; Garcia-Lamarca, M.; Triguero-Mas, M.; Cole, H.; Baró, F.; et al. Expanding the boundaries of justice in urban greening scholarship: Toward an emancipatory, antisubordination, intersectional, and relational approach. *Ann. Am. Assoc. Geogr.* **2020**, *110*, 1–27.
34. Anguelovski, I.; Shi, L.; Chu, E.; Gallagher, D.; Goh, K.; Lamb, Z.; Reeve, K.; Teicher, H. Equity impacts of urban land use planning for climate adaptation: Critical perspectives from the global north and south. *J. Plan. Educ. Res.* **2016**, *36*, 333–348. [CrossRef]
35. Harvey, D. The right to the city. *Int. J. Urban Reg. Res.* **2003**, *27*, 939–941. [CrossRef]
36. Harvey, D. *Social Justice and the City*; University of Georgia Press: Athens, Greece, 2009.
37. Marcuse, P. Reading the right to the city. *City* **2014**, *18*, 4–9. [CrossRef]
38. Lefebvre, H. The right to the city. In *Writings on Cities*; Kofman, E., Lebas, E., Eds.; Blackwell: London, UK, 1966; pp. 63–184.
39. Trundle, A. Governance and agency beyond boundaries: Climate resilience in Port Vila's peri-urban settlements In *Local Action on Climate Change: Opportunities and Constraints*; Moloney, S., Fünfgeld, H., Granberg, M., Eds.; Routledge: London, UK, 2018; pp. 35–52.
40. Baja, K.; Granberg, M. From engagement to empowerment: Climate change and resilience planning in Baltimore City. In *Local Action of Climate Change*; Moloney, S., Fünfgeld, H., Granberg, M., Eds.; Opportunities and Constraints; Routledge: London, UK, 2018; pp. 126–145.
41. Granberg, M.; Nyberg, L. Climate change adaptation, city competiveness and urban planning in the city of Karlstad, Sweden. In *Local Action of Climate Change: Opportunities and Constraints*; Moloney, S., Fünfgeld, H., Granberg, M., Eds.; Routledge: London, UK, 2018; pp. 111–125.
42. Hunt, A.; Watkiss, P. Climate change impacts and adaptation in cities: A review of the literature. *Clim. Chang.* **2011**, *104*, 13–49. [CrossRef]
43. Pelling, M. *Adaptation to Climate Change: From Resilience to Transformation*; Routledge: London, UK, 2011.
44. Granberg, M.; Glover, L. Adaptation and maladaptation in Australian national climate change policy. *J. Environ. Policy Plan.* **2014**, *16*, 147–159. [CrossRef]
45. Granberg, M.; Bosomworth, K.; Moloney, S.; Kristiansen, A.-C.; Fünfgeld, H. Can regional-scale governance and planning support transformative adaptation? *A study of two places. Sustainability* **2019**, *11*, 6978.
46. UN-Habitat. *Addressing the Most Vulnerable First: Pro-Poor Climate Action in Informal Settlements*; UN-Habitat: Nairobi, Kenya, 2018.
47. Shi, L.; Chu, E.; Anguelovski, I.; Aylett, A.; Debats, J.; Goh, K.; Schenk, T.; Seto, K.C.; Dodman, D.; Roberts, D.; et al. Roadmap towards justice in urban climate adaptation research. *Nat. Clim. Chang.* **2016**, *6*, 131–137. [CrossRef]
48. UNEP. *The Adaptation Finance Gap Report 2016*; United Nations Environment Programme (UNEP): Nairobi, Greece, 2016.
49. ÓhAiseadha, C.; Quinn, G.; Connolly, R.; Connolly, M.; Soon, W. Energy and climate policy—An evaluation of global climate change expenditure 2011–2018. *Energies* **2020**, *13*, 4839. [CrossRef]
50. Climate Policy Initiative. *Updated View of the Global Landscape of Climate Finance 2019*; Climate Policy Initiative: London, UK, 2020.
51. MDB. *Joint Report on Multilateral Development Banks' Climate Finance 2019*; European Bank for Reconstruction and Development: London, UK, 2020.
52. Taylor, P.J.; O'Brien, G.; O'Keefe, P. *Cities Demanding the Earth: A New Understanding of the Climate Emergency*; Bristol University Press: Bristol, UK, 2020.
53. Olazabal, M. An adaptation agenda for the new climate urbanism: Global insights. In *Climate Urbanism: Towards a Critical Research Agenda*; Castán Broto, V., Robin, E., While, A., Eds.; Palgrave Macmillan: London, UK, 2020; pp. 153–170.
54. Bulkeley, H.; Betsill, M.M. *Cities and Climate Change: Urban Sustainability and Global Environmental Governance*; Routledge: London, UK; New York, NY, USA, 2003.
55. Lidskog, R.; Elander, I. Addressing climate change democratically. Multi-level governance, transnational networks and governmental structures. *Sustain. Dev.* **2010**, *18*, 32–41. [CrossRef]

56. Castán Broto, V.; Robin, E.; While, A. Introduction: Climate urbanism—Towards a research agenda. In *Climate Urbanism: Towards a Critical Research Agenda*; Castán Broto, V., Robin, E., While, A., Eds.; Palgrave Macmillan: London, UK, 2020; pp. 1–11.
57. Moser, S.C. *Governance and the Art of Overcoming Barriers to Adaptation*; IHDP: Bonn, Germany, 2009.
58. Eriksen, S.H.; Nightingale, A.J.; Eakin, H. Reframing adaptation: The political nature of climate change adaptation. *Glob. Environ. Chang.* **2015**, *35*, 523–533. [CrossRef]
59. Klepp, S.; Chavez-Rodriguez, L. (Eds.) *A Critical Approach to Climate Change Adaptation: Discourses, Policies, and Practices*. Paperback ed.; Routledge: London, UK, 2020.
60. United Nations. *New Urban Agenda*; United Nations: New York, NY, USA, 2017.
61. UN FCCC. *Paris Agreement*; United Nations: NewYork, NY, USA, 2016.
62. UNISDR. *Sendai Framework for Disaster Risk Reduction 2015–2030*; UNISDR: Geneva, Switzerland, 2015.
63. United Nations. *Transforming Our World: The 2030 Agenda for Sustainable Development*; United Nations: New York, NY, USA, 2015.
64. UNOCHA. *Sustaining the Ambition: Delivering Change: Agenda for Humanity Annual Synthesis Report 2019*; United Nations: New York, NY, USA, 2019.
65. Hölscher, K. Capacities for transformative climate governance: A conceptual framework. In *Transformative Climate Governance: A Capacities Perspective to Systematise, Evaluate and Guide Climate Action*; Hölscher, K., Frantzeskaki, N., Eds.; Palgrave Macmillan: London, UK, 2020; pp. 49–96.
66. Taylor, M. *The Political Ecology of Climate Change Adaptation: Livelihoods, Agrarian Change and the Conflicts of Development*; Earthscan: London, UK, 2015.
67. Bulkeley, H.; Paterson, M.; Stripple, J. Introduction. In *Towards a Cultural Politics of Climate Change: Devices, Desires and Dissent*; Bulkeley, H., Paterson, M., Stripple, J., Eds.; Cambridge University Press: Cambridge, UK, 2016; pp. 1–23.
68. Schlosberg, D. *Defining Environmental Justice: Theories, Movements, and Nature*; Oxford University Press: Oxford, UK, 2007.
69. Schlosberg, D.; Collins, L.B. From environmental justice to climate justice: Climate change and the discourse of environmental justice. *Clim. Chang.* **2014**, *5*, 359–374.
70. Morchain, D. Rethinking the framing of climate change adaptation: Knowledge, power, and politics. In *A Critical Approach to Climate Change Adaptation: Discourses, Policies, and Practices*. Paperback ed.; Klepp, S., Chavez-Rodriguez, L., Eds.; Routledge: London, UK, 2020; pp. 55–73.
71. Taleb, N.N. *Skin in the Game: Hidden Asymmetries in Daily Life*; Random House: New York, NY, USA, 2018.
72. Anguelovski, I.; Pellow, D.N. Towards an emancipatory urban climate justice through adaptation? *Plan. Theory Pract.* **2020**, *16*, 16–21.
73. Long, J.; Rice, J.L. From sustainable urbanism to climate urbanism. *Urban Stud.* **2019**, *56*, 992–1008. [CrossRef]
74. Ferm, J.; Jones, E. Mixed-use 'regeneration' of employment land in the post-industrial city: Challenges and realities in London. *Eur. Plan. Stud.* **2016**, *24*, 1913–1936. [CrossRef]
75. Piketty, T. *Capital in the Twenty-First Century*; The Belknap Press of Harvard University Press: Cambridge, UK, 2014.
76. Molotch, H. The city as a growth machine: Toward a political economy of place. *Am. J. Sociol.* **1976**, *82*, 309–332. [CrossRef]
77. Brenner, N.; Schmid, C. Towards a new epistemology of the urban? *City* **2015**, *19*, 151–182. [CrossRef]
78. Soja, E.W. *Seeking Spatial Justice*; University of Minnesota Press: Minneapolis, Minnesota, 2010.
79. Chatterton, P. Seeking the urban common: Furthering the debate on spatial justice. *City* **2010**, *14*, 625–628. [CrossRef]
80. Glacken, C.J. *Traces of the Rhodian Shore: Nature and Culture in Western Thought from Ancient Times to the End of the Eighteenth Century*; University of California Press: Berkeley, CA, USA, 1967.
81. Engels, F. *Condition of the Working Class in England*; Oxford University Press: Oxford, UK, 1845/1993.
82. Fainstein, S.S. The just city. *Int. J. Urban Sci.* **2014**, *18*, 1–18. [CrossRef]
83. Lefebvre, H. The Right to the City. In *Writings on Cities*; Kofman, E., Lebas, E., Eds.; Blackwell: London, UK, 1967; pp. 63–194.
84. Moloney, S.; Fünfgeld, H.; Granberg, M. Climate change responses from the global to the local scale: An overview. In *Local Action on Climate Change: Opportunities and Constraints*; Moloney, S., Fünfgeld, H., Granberg, M., Eds.; Routledge: London, UK, 2018; pp. 1–16.
85. Bulkeley, H.; Tuts, R. Understanding urban vulnerability, adaptation and resilience in the context of climate change. *Local Environ.* **2013**, *18*, 646–662. [CrossRef]
86. Hoppe, T.; van der Vegt, A.; Stegmaier, P. Presenting a framework to analyze local climate policy and action in small and medium-sized cities. *Sustainability* **2016**, *8*, 847. [CrossRef]
87. Sovacool, B.K.; Brown, M.A. Scaling the policy response to climate change. *Policy Soc.* **2009**, *27*, 317–328. [CrossRef]
88. Bulkeley, H. *Accomplishing Climate Governance*; Cambridge University Press: New York, NY, USA, 2016.
89. Klepp, S.; Chavez-Rodriguez, L. Governing climate change: The power of adaptation discourses, policies, and practices. In *A Critical Approach to Climate Change Adaptation: Discourses, Policies, and Practices, Paperback ed.*; Klepp, S., Chavez-Rodriguez, L., Eds.; Routledge: London, UK, 2020; pp. 3–34.
90. Hölscher, K.; Frantzeskaki, N. A transformative perspective on climate change and climate governance. In *Transformative Climate Governance: A Capacities Perspective to Systematise, Evaluate and Guide Climate Action*; Hölscher, K., Frantzeskaki, N., Eds.; Palgrave Macmillan: London, UK, 2020; pp. 3–48.
91. Jasanoff, S. A new climate for society. *Theory Cult. Soc.* **2010**, *27*, 233–253. [CrossRef]

92. Bulkeley, H.; Paterson, M.; Stripple, J. *Towards a Cultural Politics of Climate Change: Devices, Desires, and Dissent*; Cambridge University Press: Cambridge, UK, 2016.
93. Granberg, M.; Elander, I. Local governance and climate change: Reflections on the Swedish experience. *Local Environ.* **2007**, *12*, 537–548. [CrossRef]
94. Bulkeley, H.; Castán Broto, V.C.; Maassen, A. Governing urban low carbon transitions. In *Cities and Low Carbon Transitions*; Broto, V.C., Bulkeley, H., Hodson, M., Marvin, S., Eds.; Routledge: London, UK, 2013; pp. 29–41.
95. Granberg, M. Strong local government moving to the market. In *Rethinking Urban Transitions: Politics in the Low Carbon City*; Luque-Ayala, A., Marvin, S., Bulkeley, H., Eds.; Routledge: London, UK, 2018; pp. 129–145.
96. Bulkeley, H.; Edwards, G.A.S.; Fuller, S. Contesting climate justice in the city: Examining politics and practice in urban climate change experiments. *Glob. Environ. Chang.* **2014**, *25*, 31–40. [CrossRef]
97. Schlosberg, D. Climate Justice and Capabilities: A Framework for Adaptation Policy. *Ethics Int. Aff.* **2012**, *26*, 445–461. [CrossRef]
98. Gardiner SMClimate justice. *The Oxford Handbook on Climate Change and Society2011*; Dryzek, J.S., Norgaard, R.B., Schlosberg, D., Eds.; Oxford University Press: New York, NY, USA, 2011; pp. 309–322.
99. Quigley, W.P. What Katrina revealed. *Harv. Law Policy Rev.* **2008**, *2*, 361–384.
100. Trundle, A.; McEvoy, D. Climate Change Vulnerability Assessment: Greater Port Vila. In *Cities and Climate Change Initiative*; UN-Habitat: Fukuoka, Japan, 2015.
101. Butcher-Gollach, C. Planning, the urban poor and climate change in Small Island Developing States (SIDS): Unmitigated disaster or inclusive adaptation? *Int. Dev. Plan. Rev.* **2015**, *37*, 225–248. [CrossRef]
102. Trundle, A.; Barth, B.; Mcevoy, D. Leveraging endogenous climate resilience: Urban adaptation in Pacific Small Island Developing States. *Environ. Urban.* **2019**, *31*, 53–74. [CrossRef]
103. Mitchell, D.; McEvoy, D. *Land Tenure and Climate Variability*; UN-Habitat: Nairobi, Kenya, 2019.
104. Biehl, E.; Buzogany, S.; Baja, K.; Neff, R.A. Planning for a resilient urban food system: A case study from Baltimore City, Maryland. *J. Agric. Food Syst. Community Dev.* **2018**, *8*, 39–53. [CrossRef]
105. Teodoro, J.D.; Nairn, B. Understanding the knowledge and data landscape of climate change impacts and adaptation in the Chesapeake Bay Region: A systematic review. *Climate* **2020**, *8*, 58. [CrossRef]
106. Ntelekos, A.A.; Oppenheimer, M.; Smith, J.A.; Miller, A.J. Urbanization, climate change and flood policy in the United States. *Clim. Chang.* **2010**, *103*, 597–616. [CrossRef]
107. Sarzynski, A. Multi-level governance, civic capacity, and overcoming the climate change "adaptation deficit" in Baltimore, Maryland. In *Climate Change in Cities: Innovations in Multi-Level Governance*; Hughes, S., Chu, E.K., Mason, S.G., Eds.; Springer: Cham, Switzerland, 2018; pp. 97–120.
108. Huang, G.; Zhou, W.; Cadenasso, M.L. Is everyone hot in the city? Spatial pattern of land surface temperatures, land cover and neighborhood socioeconomic characteristics in Baltimore, MD. *J. Environ. Manag.* **2011**, *92*, 1753–1759. [CrossRef]
109. Hrelja, R.; Hjerpe, M.; Storbjörk, S. Creating transformative force? The role of spatial planning in climate change transitions towards sustainable transportation. *J. Environ. Policy Plan.* **2015**, *17*, 617–635. [CrossRef]
110. Storbjörk, S.; Hjerpe, M.; Isaksson, K. We cannot be at the forefront, changing society: Exploring how Swedish property developers respond to climate change in urban planning. *J. Environ. Policy Plan.* **2018**, *20*, 81–95. [CrossRef]
111. Engström, G. När Vänerns och Klarälvens vatten möts på Karlstads torg kommer staden att gå under. In *Klarälven*; Ibsen, H., Svensson, E., Nyberg, L., Eds.; Karlstad University Press: Karlstad, Sweden, 2011; pp. 93–105.
112. Granberg, M.; Nyberg, L.; Modh, L.-E. Understanding the local policy context of risk management: Competitiveness and adaptation to climate risks in the city of Karlstad, Sweden. *Risk Manag.* **2016**, *18*, 26–46. [CrossRef]
113. Moloney, S.; Fünfgeld, H.; Granberg, M. Towards transformative action: Learning from local experiences and contexts. In *Local Action on Climate Change: Opportunities and Constraints*; Moloney, S., Fünfgeld, H., Granberg, M., Eds.; Routledge: London, UK, 2018; pp. 146–156.
114. Glass, R. *London, Aspects of Change*; MacGibbon & Kee: London, UK, 1964.
115. Lees, L.; Phillips, M. (Eds.) *Handbook of Gentrification Studies*; Edward Elgar: Cheltenham, UK, 2018.
116. Rawls, J. *A Theory of Justice*; The Belknap Press of Harvard University Press: Cambridge, UK; London, UK, 1971.
117. Stone, C.N. City politics and economic development: Political economy perspectives. *J. Politics* **1984**, *46*, 286–299. [CrossRef]
118. Long, J.; Levenda, A. Climate urbanism and the implications for climate apartheid. In *Climate Urbanism: Towards a Critical Research Agenda*; Castán Broto, V., Robin, E., While, A., Eds.; Palgrave Macmillan: London, UK, 2020; pp. 31–49.
119. Davies, J.; Ziervogel, G. Learning by doing: Lessons from the co-production of three South African municipal climate adaptation plans. In *Local Action on Climate Change: Opportunities and Constraints*; Moloney, S., Fünfgeld, H., Granberg, M., Eds.; Routledge: London, UK, 2018; pp. 53–71.
120. Ciplet, D.; Roberts, J.T. Climate change and the transition to neoliberal environmental governance. *Glob. Environ. Chang.* **2017**, *46*, 148–156. [CrossRef]
121. Robbins, P. *Political Ecology: A Critical Introduction*; Wiley-Blackwell: Chichester, UK, 2012.
122. Watts, M. Political ecology. In *A Companion to Economic Geography*; Sheppard, E., Barnes, T.J., Eds.; Blackwell Publishing: London, UK, 2000; pp. 257–274.

123. Paulsson, S.; Gezon, L.L.; Watts, M. Locating the political in political ecology: An introduction. *Hum. Organ.* **2003**, *62*, 205–217. [CrossRef]
124. O'Brien, K.; Eriksen, S.; Nygaard, L.P.; Schjolden, A. Why different interpretations of vulnerability matter in climate change discourses. *Clim. Policy* **2007**, *7*, 73–88. [CrossRef]

Article

Conditions and Constrains for Reflexive Governance of Industrial Risks: The Case of the South Durban Industrial Basin, South Africa

Llewellyn Leonard [1,*] and Rolf Lidskog [2]

1 Department of Environmental Sciences, College of Agriculture and Environmental Sciences, University of South Africa (UNISA), Johannesburg 1709, South Africa
2 Environmental Sociology Section, School of Humanities, Education and Social Sciences, Örebro University, SE-701 82 Örebro, Sweden; Rolf.Lidskog@oru.se
* Correspondence: llewel@unisa.ac.za; Tel.: +27-792442087

Citation: Leonard, L.; Lidskog, R. Conditions and Constrains for Reflexive Governance of Industrial Risks: The Case of the South Durban Industrial Basin, South Africa. *Sustainability* **2021**, *13*, 5679. https://doi.org/10.3390/su13105679

Academic Editors: Anna Visvizi and Marc A. Rosen

Received: 13 April 2021
Accepted: 14 May 2021
Published: 19 May 2021

Publisher's Note: MDPI stays neutral with regard to jurisdictional claims in published maps and institutional affiliations.

Copyright: © 2021 by the authors. Licensee MDPI, Basel, Switzerland. This article is an open access article distributed under the terms and conditions of the Creative Commons Attribution (CC BY) license (https:// creativecommons.org/licenses/by/ 4.0/).

Abstract: Within sustainability development paradigms, state governance is considered important in interventions to address risks produced by the industrial society. However, there is largely a lack of understanding, especially in the Global South, about the nature and workings of the governance institutions necessary to tackle risks effectively. Reflexive governance, as a new mode of governance, has been developed as a way to be more inclusive and more reflexive and respond to complex risks. Conversely, there is limited scholarly work that has examined the theoretical and empirical foundations of this governance approach, especially how it may unfold in the Global South. This paper explores the conditions and constrains for reflexive governance in a particular case: that of the South Durban Industrial Basin. South Durban is one of the most polluted regions in southern Africa and has been the most active industrial site of contention between local residents and industry and government during apartheid and into the new democracy. Empirical analysis found a number of constrains involved in enabling reflexive governance. It also found that a close alliance between government and industry to promote economic development has overshadowed social and environmental protection. Reflexive governance practitioners need to be cognisant of its applicability across diverse geographic settings and beyond western notions of reflexive governance.

Keywords: deliberative democracy; ecological reflexivity; reflexive governance; participation; regulation; risk; transparency

1. Introduction

The urgent need for sustainable development raises issues of governance, since sustainability goals are subjected to heterogeneous perceptions and interest [1]. To reach necessary transformative change, there is a need to both properly understand this situation as well as find ways to manage it that are both politically legitimate and relevent to the environment. Although new modes of governance, such as reflexive governance, can increase participation and deliberation across industry, government and civil society sectors, thereby providing more legitimate decision-making, there is limited scholarly work that has examined the theoretical and empirical foundations of this assumption [2], let alone in the Global South. This includes the structures, power relations and actions that may hinder the emergence of reflexive governance [3]. There is apprehension about the political implications of reflexive governance since its designs engage with real-world political contexts, which affect their workings and may weaken their efficiency. Additionally, there may also be worries caused by the democratic legitimacy of reflexive governance designs and the uncertain relationships with establishments of representative democracy [4]. There are further concerns that reflexive governance emerged from the 1990s and developed in an era in which neoliberalism was the dominant political discourse, with repeated efforts to reduce the power of the nation state in favour of industry's self-regulation [5]. Swyngedouw and Kaika (2014) and Dagkas and Tsoukala (2011) note that neo-liberalisation makes it difficult for vulnerable groups to have equal access to good-quality environmental resources, and for procedural quality in decision-making to occur [6,7]. However, linked to neoliberalism, Rosenau notes that government institutions can evolve in such a way as to be minimally dependent on hierarchical, command-based arrangements (i.e., industrial deregulation and self-regulation; loss of governance functions by the state) [8]. Nevertheless, the point of this paper is not to engage in complex debates about the ills of neoliberalism or to provide solutions to neoliberalism. The authors believe that the solutions to neoliberalism must evolve through genuine discussions between civil society, government and industry on moving towards sustainable development. As Luna (2015) states, the movement away from neoliberalism is about having a discussion about the kind of development we want for our future, how basic needs will be secured for everyone, moving away from inequality and how these goals will be achieved [9]. However, there is no doubt that reflexive governance will be important in these discussions, and there is a need to investigate how reflexive governance may be strengthened.

Reflexive governance may face a number of challenges such as how to treat and deal with the state's power, responsibility, boundaries, the withering manner of the state, the problem between state management and state governance, and the problem of long-term coexistence and positive interaction between state and society, etc. [4,10]. Modern approaches to reflexive governance may thus aspire technocratic approaches to governance, which give rise to institutions that yield instability, whilst ignoring environmental externality impacts [11] and lack the capacity to co-ordinate collective action due to non-hierarchical forms of governance [12]. The question is whether reflexive governance may be a hybrid mode of governance, interpenetrated by other modes, or if it exists alongside and/or in competition with them [13]. Reflexive governance is, therefore, not straightforward and involves managing a plethora of contestations over sustainability and acknowledging that legitimacy is negotiated [14]. Limited research has explicitly investigated the potential for reflexivity to assist in understanding the politics of human–environmental impacts [11] and how reflexive governance unfolds or may potentially spiral into poor governance and risk ignoring or fragmenting divergent views [12] with reflexivity as one of the tenets of reflexive governance [15]. A major shortcoming of the existing literature on reflexivity is that the distinction between what reflexive governance is and what enables and/or hinders it is unclear [11]. If neoliberalism may influence reflexive governance approaches, then how may reflexive governance principles be safeguarded to ensure that they do not spiral into a technocratic approach or become paternalistic, thereby perpetuating risks?

Within this context, the aim of this paper is to explore the conditions and constrains of reflexive governance. Of particular importance is to examine the application of reflexive governance, its political implications, and shortcomings in terms of addressing industrial risks, using the case of the South Durban Industrial Basin (SDIB) in KwaZulu-Natal, South Africa. Whilst there is largely a lack of understanding, especially in the Global South, about the nature and workings of the governance institutions necessary to tackle risks effectively, and which Southern countries exhibit different and perhaps more severe technical, financial and capacity constraints to Northern countries, it is important to explore the applicability of reflexive governance across diverse geographic settings, especially at the local level. As Guay also highlights, the connections between local governments and global processes receive limited attention in the political economy, despite local governance forays into the foreign policy realm having important implications for governance and policy-making generally [16]. The SDIB is considered to be one of the most contaminated regions in southern Africa and is declared a pollution hotspot. This will include how the state manages the plethora of contestations regarding risks. The geographic setting of local communities which have been historically exposed to petrochemical industries and industrial risks has not transformed since the introduction of democracy [17], making it of interest to investigate to what extent, and in what way, reflexive governance approaches has been employed. This paper consists of seven sections, including this introduction. The Section 2 provides some background to South Africa and reflexive governance. The Section 3 explores the literature on reflexive governance, participation and deliberative democracy. The section also outlines supplementary approaches to reflexive governance, such as 'value reflexive governance' and 'ecological reflexivity' approaches. The Section 4 explores the literature on an 'enabling' reflexive governance approach. The Section 5 outlines the study, the case description, the empirical material and the applied method. The Section 6 presents the results. The Section 7, which is the conclusion, evaluates the findings in relation to the literature and discusses the wider implications for an enabling a reflexive governance approach.

2. South Africa and Reflexive Governance

South African achieved democracy in 1994, an era when neoliberal ideology was globally dominant [18,19], and during a period when reflexive governance emerged. As a newly democratised nation, and in line with reflexive governance principles, the country developed a democratic constitution and a democratic parliament. For example, the 1996 Constitution of the Republic of South Africa makes provisions for the right to a healthy environment. It also bestows the right to have the environment protected by the government, who must prevent pollution and degradation, and ensure ecological sustainable development. It makes provisions for the participation of citizens in decision-making processes, democracy and accountability, the separation of powers and cooperative governance, and the decentralization of power [20]. Various policies and regulations have also been developed in line with the constitution to reduce and eliminate environmental and social risks in society. An example is the 1998 National Waste Management Strategy (NWMS), designed to ensure the good health of the people and the quality of the environment by the implementation of 'cleaner production' to increase the eco-efficiency of industrial processes, as well as to implement 'cleaner technologies' to reduce pollution and industrial risks in society [21]. The 1998 National Environmental Management Act (NEMA) emphasises that people's needs must be put at the forefront when matters of environmental management are considered, and makes provisions for the promotion of the participation of all interested and affected parties in environmental governance [22]. 'Reflexive governance' approaches to deliberation and participation between a variety of stakeholders has, therefore, been considered critical for implementing processes to address societal risks, with civil society considered an important aspect of the new, inclusive 'democratic' societies. These act to ensure the human right to a healthy and just environment. Since its establishment of democracy, South Africa has implemented regulations that provide guiding principles for a 'reflexive governance' approach to an inclusive society so as to increase participation and deliberation between the government, industry and civil society.

Unfortunately, despite the progress in enabling the supportive governance policy frameworks and the values to preventing and managing environmental risk, the implementation of reflexive governance principles has been limited [23]. As Malherbe and Segal (2001) note, although South African legislation has attempted to sharpen corporate accountability for corporate actions post-1994, government institutions have not actively and publicly monitored corporate governance [24]. Due to the government engaging in a macroeconomic neoliberal model since its democracy, it has concentrated on expanding industrial modes of production [25–27], with the logic of wealth production dominating the logic of risk alleviation, which contributes to increased industrial risks in society [28]. The state has also recognised, to some degree, its own incapacity to regulate effectively, with enforcement being inadequate or mechanistic, addressing the tail end rather than holistic approaches to address industrial risks. This has stifled technological innovation, stressing supply-side solutions rather than behavioural change on the part of industry [29]. Additionally, the anti-apartheid struggle should have impressed on the new government the need to incorporate citizens' interests into decision-making processes; however, participation mechanisms have not enhanced participatory governance. Participatory mechanisms have been biased towards citizens with the capacity and resources to organise, who would be able to bring their concerns to government attention without these mechanisms [30]. Despite the general literature on state governance deficiencies, there is a lack of empirical evidence on how reflexive governance may be obstructed towards addressing industrial risks.

3. Reflexive Governance, Participation and Deliberative Democracy

The concept of reflexivity arose due to the industrial society producing unforeseen and unintended side effects as a result of an unlimited faith in science, bureaucracy and instrumental rationality [11,13,31–33]. In Beck's periodization of social change, simple modernity is associated with the development of industrial society, whilst the new, reflexive modernity is associated with the emergence of the risk society, in which progress can turn into self-destruction [24,31]. Fuelled by technical disasters, the scientific capacity to determine risks and propose viable ways to handle them has been questioned [34,35], as well as industrial expertise [15] concerning its interest and ability to shape structural change in society and technology [1]. Within the sustainability development paradigms, state governance is, therefore, considered important in interventions to address unintended side effects and manage risks. The theory of reflexive modernization does not include the demise of the state, which simultaneously remains both the agent and the subject of change. Although aspects of the nation-state have been undermined, the nation-state still retains a considerable role in the governance process. 'Governance', in turn, is recast as a mechanism for managing today's pervasive uncertainty. Reflexive modernisation allows for the recasting of 'governance' as a necessary, yet contingent, mechanism of managing uncertainty in contemporary societies [36].

Although 'governance' has diverse interpretations [37,38], modern approaches to governance are generally understood as the inclusion of the non-state stakeholders in decision-making [39–42] and emphasis of accountability, transparency, fairness, rule of law and ethical considerations by the state [13,43], whilst not relying on technocratic and bureaucratic processes to manage developmental and policy processes [1,29,44]. This collective understanding of modern governance can be grouped together under reflexive governance [4]. Reflexive governance, as a new mode of governance, is viewed as organising a response to the risks by replacing traditional, hierarchical and deterministic governance approaches with more reflexive, flexible and interactive ones, which draw on diverse knowledge systems [2,12,13,45]. Despite this understanding of reflexive governance, there is largely a lack of understanding about the nature and workings of the governance institutions that are necessary to effectively enable reflexive governance in society, so as to tackle industrial risks [3,46,47].

Participation and deliberation are central to reflexive governance and democracy, and to tackling development challenges [3], with reflexivity also associated with the principle of participation [14]. The concepts of governance and participation are interrelated, as governance is difficult to achieve if participation is insufficient. An essential component of good governance is the ability to enable citizens to express their views, and to act on those views, facilitated through participation [48]; the more deliberate the process, the more reflexive governance is [3]. When in-depth information is not disseminated to citizens, participatory and deliberation mechanisms may be ineffective [49]. Formal participatory assemblies may sometimes be geared towards 'domesticating' and undermining the legitimacy of groups who choose to engage critically with local governments (and industry). This has the potential to revert back to 'first generation' governance (i.e., traditional state-centred and technocratic regulation) and move away from the actual principles and values of reflexive governance.

For example, Wesselink et al. (2011) noted that impediments to participation may occur when environmental policies are not aligned with other policies and when economic interests prevail over environmental issues. Thus, 'participation fatigue' can occur, which is the failed embedding of new participatory governance in a bureaucratic structure that is not receptive to input from other stakeholders (e.g., civil society) [44]. To work towards sustainable development, the encouragement of knowledge inputs and participation from across society is not just an instrumental imperative, but an ethical imperative, since it is only on the basis of interactive governance that it is possible to elaborate a development trajectory that reflects the fundamental needs of society at large [14]. When linking reflexive governance with 'deliberative democracy', participants can debate the various issues in a careful and reasonable fashion for democratic legitimacy to occur. Only after genuine discussions occur, can decisions be made. In this sense, the deliberative aspect corresponds to a collective process of reflection and analysis, permeated by the discourse that precedes the decision. However, despite the principles of interactive governance within a reflexive governance approach, there is no guarantee that government (or industry) will genuinely apply these principles.

Therefore, it is important to distinguish between genuine processes of participation and deliberation, as opposed to more tokenistic ones. It is useful to draw on the ladder of participation, as presented by Arnstein (1969), which is still useful in understanding the different types of participation. These are grouped into 'non-participation', 'tokenism', and 'citizen power'. With 'non-participation' (i.e., manipulation and therapy), the objective is to gain support for decisions which are already made. 'Tokenism', namely, informing, consultation and placation, allows citizens to express their views but with no assurance that citizens' concerns will be taken into account. 'Citizen power' (namely, partnership, delegated power and citizen control) results in an increase in citizens' decision-making powers. 'Partnerships' allow for power to be equally shared among citizens and power-holders. Regarding 'citizen control', Arnstein notes that, although citizens demand a degree of power (for example, governing of a program or institution), a Model City cannot meet the criteria of citizen control, since final approval power rests with the city authority. Nevertheless, citizen empowerment suggests that direct democracy (the participation of citizens in decision-making) needs to be established on a 'partnership' basis, with citizens treated as equal partners in development and decision-making processes [50].

4. Towards an 'Enabling' Reflexive Governance Approach

Reflexive governance, under the auspices of the 'next generation' environmental regulations, has been proposed as a means to overcome various insufficiencies associated with the 'first generation' environmental regulation linked with direct command and control state regulation and failure to nurture contextualised learning [2,4]. In relation to the 'old' and 'new' modes of governance, it is useful to draw on the various distinctions of governance, as noted by Stirling (2006) [51], which refer to unreflective, reflective and reflexive governance approaches. Unreflective governance denotes limited instrumental driven decision-making processes, whilst reflective governance involves more critical attempts to manage side effects and garner a multitude of perspectives to implement the best policy. Reflexive governance is about engaging with a variety of social actors, rather than eliminating ambivalence [2,14,52]. It seeks to explore how ambivalence is incorporated into reflexive approaches through governance, with participatory processes of various forms widely advocated to understand social change [4,52]. This is, in most instances, geared towards the construction of collective and consensual visions of what a more sustainable socio-technical system might entail [53]. Reflexive governance emphasises the participatory approaches in goal formulation and strategies of development for governance [54]. Since ambivalences of sustainability are emergent, reflexive governance is assumed to have the distinct feature of continually monitoring, feeding back and adjusting as a means of handling these interdependencies and the unpredictability of systemic change [52].

Considering that reflexive governance may become compromised within a neoliberal and bureaucratic framework, how can reflexive governance be robustly implemented and safeguarded? A clearer and enabling conception of reflexive governance can be facilitated through concepts such as 'value reflexive governance' and 'ecological reflexivity'. The idea is not to posit one concept over the other, but rather to have them work together, since they are based on particular values. Meisch et al. (2012) notes that, as opposed to reflective governance, 'value reflexive governance' emphasises the values of good governance norms, where values become the guiding imperatives, which offer more transparent and inclusive governance by not only making it imperative for more social actors to express their values, but ensuring that those values materialise into actions [55]. According to Smith and Stirling (2007), reflexive governance is also about social actors changing values in response to scrutiny [56]. Meisch et al. (2012) states that value reflexive governance is sensitive to participants' values in governance processes and develops specific solutions to problems. Unfortunately, there is no set of particular philosophical tools to guide value discourses, and there are diverse social and institutional contexts which maybe unique to particular environments [55]. Reflecting on ways to develop procedural arrangements that work with power and conflict, to ensure inclusive participation, equality among participants and open communication in a process of experimentation and learning, is necessary to cope with ambivalence [4].

According to Weiland (2012), there is a need to collectively develop the *instruments* and *procedures* that allow for the productive use of diversity. It seems that deliberation between all social actors is key to ensuring the development of the 'value' instruments and procedures for enabling reflexive governance [12]. It is also important to explore what determines reflexive governance. As Meadowcroft and Steurer (2018) note, reflexive governance is not straightforward [14]. As highlighted, modern approaches to reflexive governance may spiral into technocratic approaches to governance and overlook environmental externality impacts. Pickering (2019) attempts to solve this problem by further proposing 'ecological reflexivity', understood as, 'the capacity of an entity (for example, an agent, structure, or process) to recognise its impacts on social-ecological systems and vice-versa; rethink its core values and practices in this light, and respond accordingly by transforming its values and practices [11]. Therefore, this goes beyond policy processes initiated by government and moves from unconscious reaction to conscious and cognitive effort. Figure 1 shows the three components and their signs of ecological reflexivity. According to Pickering, in order to qualify as minimal reflexivity, institutions (for example, government and industry) must show all three components of reflexivity to some degree. This understanding can assist in detecting shortfalls in reflexivity and unveil non-reflexive or reflexive institutions. For example, an agent or structure will be considered non-reflexive if it recognises a problem and fails to rethink activities such as the government's lack of enforcement of environmental regulations, leading to the continual pollution of society and the environment by industrial processes.

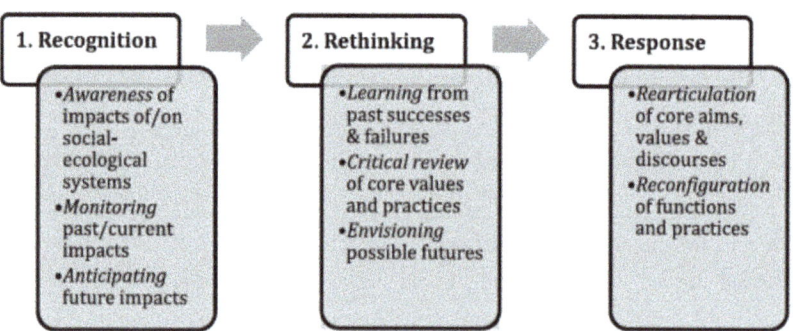

Figure 1. Components and signs of ecological reflexivity: Source: Pickering (2019).

The combination of value-reflexive governance and ecological reflexivity in reflexive governance approaches can ensure that there is a more explicit emphasis on the formulation of participants' values to formulate actions, and that all stakeholders can develop *instruments* and *procedures* that will enable the productive use of diversity with feedback on decisions made. Social actors are also enabled to recognise their impact on social-ecological systems and must be reflexive in rethinking practices, thereby transforming values and practices. Such an approach can entail more open, transparent and inclusive governance. As Meisch et al. (2012 notes, it is important that values are turned into actions by developing specific solutions to the problems enabled by the formulation of values, instruments and procedures [55]. The key, however, is that the participation and deliberation encouraged by the government ensures the decentralisation of power to citizens, so that all stakeholders have an equal chance to induce positive change [57] and ensure equity and fairness in the rule and application of law [38,43].

5. Materials and Methods

5.1. The Studied Case

The South Durban Industrial Basin (SDIB) in KwaZulu-Natal, South Africa is considered one of the most contaminated regions in southern Africa. The SDIB is home to two of South Africa's four oil refineries, Africa's foremost chemical storage facility, and over 180 smokestack industries are located in the SDIB [58]. The region is an example of an industrial hub that includes residential areas situated next to heavy industries [18]. The apartheid impression to formulate a deliberate industrial region was efficaciously implemented by the early 1970s; this region became home to 70 percent of Durban's industrial activity [59].

A brief background about South Durban is that in 1994, with the ushering in of democracy, the community established the South Durban Community Environmental Alliance (SDCEA) to tackle the issue of pollution within South Durban. (The SDCEA is made-up of 16 affiliate organisations and has been active since its formation. It makes no profit and exists solely for the benefit of the people it represents. The Alliance is vocal and active in lobbying, reporting and researching industrial incidents and accidents in South Durban [60].) According to Reid and D'Sa (2005), SDCEA was formed to link local concerns across racial boundaries in order to respond systematically to pollution issues [61]. As was the case during apartheid, South Durban communities continue to endure the environmental, health, and socio-economic costs of pollution from adjacent industries [18], with the most frequent perpetrators of atmospheric pollution incidents being the oil refineries, notably Engen Petroleum and the South African Petroleum Refinery (SAPREF) [60]. Health impacts in South Durban have been a concern, with asthma rates documented at double the global average [62], and the incidence of leukaemia noted to be up to 24 times higher than the national average [63]. A 2002 medical study carried out by a team of medical researchers at the local Settlers Primary School bordering the Engen refinery found that 52 percent of learners suffered from severe asthma, and 11 percent of learners experienced moderate to severe persistent asthma. It was further found that children in South Durban were much more likely to suffer from chest complaints than children from other parts of Durban [64]. The SDCEA also recorded a total of 55 major industrial incidents in South Durban from 2000 to 2016 [65,66]. In 2018, the South Durban basin was declared a pollution hotspot, according to the provincial government's Environment Outlook Report [67].

5.2. Data Collection

For the qualitative research methodology employed for this study, primary material was collected by one of the authors in June and July 2019. Semi-structured interview guides were used to collect data from key social actors (namely, local government, external civil society scientific experts supporting community groups, local residents, and community-based organisations (CBOs)). A purposive sampling design was employed during fieldwork, where the researcher's judgement determines who can provide the best information to achieve the objectives of the study. A snowballing technique was also employed, as informants referred the researcher to other informants for interview. A total of seven interviews were conducted, of which five interviews are reported in this study. One interview was secured with an anonymous local government official. Two interviews were secured with academics from the University of KwaZulu-Natal, dealing with health studies and industrial expansion in South Durban, respectively. One interview was secured with an ex-local government employee (now working in the water industry in the same area), who was previously responsible for air pollution in South Durban, and another interview was conducted with the SDCEA leadership. Unfortunately, numerous attempts to secure interviews with several major South Durban industries proved futile. Additionally, two key government officials directly responsible for pollution issues and air-quality monitoring in South Durban did not respond to interview invitations. This has implications for the data analysis, as the views of two important social actors have not been possible to obtain.

This paper also draws on fieldwork conducted in South Durban in July 2017, which explored governance as part of a study that looked at 'toxic tourism' for environmental justice [68]. Semi-structured interviews were conducted with several members of the local SDCEA involved with various environmental projects in South Durban, ranging from industrial pollution to climate change. From this study, two interviews are used that particularly concern governance issue.

Documents are used as secondary sources when necessary to identify relationships or patterns regarding the interview content and as a way to interrogate and verify what has been said. These include the 2007 South Durban Basin Multi-Point Plan (MPP) Case Study Report [69]; the 2017/2018 eThekwini Municipality Annual Report [70]; a 2019 Independent Online (IOL) media article [71]; the 2020 National Air Quality Indicator—Monthly data report [72]; a 2018 Mail and Guardian media article [73]; a 2019 Africa News Agency media article [74]; the 2018 South African Petroleum Industry Association (SAPIA) Annual Report [75]; and the 2018 Engen Refinery Integrated Report [76].

5.3. Data Analysis

For the data analysis, grounded theory and open coding were employed to identify similar emerging themes across the interviews (namely, scientific experts and co-creation of knowledge, risk communication, trust and transparency, participation, governance, civil society resources and fragmentation). Grounded theory and open coding primarily involve taking the data apart (i.e., interviews) and examining the parts for differences and similarities. Codes are clustered together to form categories (i.e., themes) [77]. This article focuses on two main themes (namely, poor governance skills and transparency when addressing industrial risks, and the lack of participatory governance and deliberation), which are discussed below. There are links and overlaps between the themes.

5.4. Informed Consent

All subjects gave their informed consent for inclusion before they participated in the study. The study was conducted in accordance with the Declaration of Helsinki, and the protocol for the 2019 field work was approved by the University of South Africa, College Research Ethics Committee (Project identification REC 170616-051). For the fieldwork of 2017, the ethical standards for academic research were followed. It was, however, not possible to obtain ethical clearance from the University of South Africa, because the study was based at another university which, at that time, had no ethics committee in place.

6. Results

In the following, we will analyse to what extent reflexive governance was applied to address the industrial risks in the SDIB. In particular, the hindrances in applying this approach will be investigated. This section presents two main themes (i.e., lack of governance skills and transparency to address industrial risks, and lack of participatory governance and deliberation).

6.1. Lack of Governance Skills and Transparency to Address Industrial Risks

Poor governance, since 2010, has resulted in the local government not maintaining air-monitoring equipment to collect air-quality data so as to address potential industrial risks in South Durban, as per the previous MPP. This, in turn, had implications for the government not sharing air-quality information with civil society in order to make informed decisions about industrial development issues, with implications for transparency within the new reflexive governance democratic dispensation. The MMP was launched after the dawn of democracy, in November 2000, and aimed to provide an improved and collective decision-making structure for air pollution management at the local government level, reduce air pollution to meet health-based air-quality standards, and improve the quality of life for the local community. Some of the key achievements of the MMP were an improved air-quality monitoring network with integrated data transfer and storage, the extension of sampling to other pollutants such as Benzene, Toluene, Ethyl-benzene and Xylene, and a reflexive government multi-stakeholder approach, with the government, community and industry working jointly [69]. The ex-government employee, Informant A, working for a water purification company in South Durban, who was previously employed as a local government official responsible for pollution control and programme manager of the previous MPP, noted the deterioration of governance in combatting industrial risks in South Durban since the MPP ended:

'There have been gains and changes [to combat industrial risks since the turn of democracy] but I am a bit disappointed in the maintenance of that current domain ... The Multi-Point Plan was funding from industry, funding from government and there was a true multi-stakeholder nature of a democratic government and industry coming on board and led by the top teams. The MPP delivered as it enabled the science to prevail with monitoring and good data and allowing a health study to take place. A first successful health study that is also documented and peer-reviewed. Industry started investing at that time millions of Rands [South African currency] to address pollution. So under the MPP we have seen pollution coming down to within levels and now it is best practice from industry ... But what is disappointing and since I left government, there has been a disinterest [by local government] in maintaining the [air] monitoring stations ... I complained [to local government] about an incident ... and it took many weeks for them to respond, but an inadequate response ... You have to look at the stakeholder approach to handle the issue, making sense of it to make a solution. So that is missing ... They [government] will probably give you lots of excuses why things can't happen ... ' (Informant A, 19 June 2019).

The local SDCEA CBO further noted the lack of proper reflexive governance to maintain air-monitoring equipment in South Durban, including the lack of transparency and communication to civil society surrounding how the data were collected by the government. Informant B, a SDCEA Environmental Project Officer—responsible for Development, Infrastructure and Climate Change—noted for the lack of governance to monitor air pollution:

' ... We have the city with the latest [air monitoring] equipment ... that will give you [an] exact reading of what's emitted out [by industry] and what volume. [It has been] non-functional since 2013, yet to date, two, three weeks ago, they have claimed that no, we've only stopped receiving information from our monitoring stations from 2016. When we asked them about that, we didn't know that and one of their [government] portfolio committee members in Johannesburg stated that they have not received any information to date since 2013. So, these are the key things that are displaying themselves in our communities and government is not acting on our behalf ... ' (Informant B, 10 July 2017).

The above suggests an unreflective governance approach by the municipality in not organising a response to potential risks and not displaying accountability and transparency towards external stakeholders. In an article published by the Independent Online (IOL) media in 2019, it was reported that many civil society groups complained that local government had not supplied and engaged with them regarding air-pollution monitoring information for many years, as the monitoring stations were non-functional [71]. Reference to the National Air Quality Indicator Monthly data report for the KwaZulu Natal Province, April 2020 [72], stipulates that the 2004 National Air Quality Act requires the establishment of national standards for municipalities and provinces to monitor ambient air quality to report compliance with ambient air quality standards. To this end, different spheres of government have invested in continuous air-pollution monitoring hardware to meet this objective. The information secured from these stations is used to develop the National Air Quality Indicator (NAQI), based on an annual measure of prevalent pollutants. One of the purposes of the NAQI is to provide an evidence-based approach in presenting and measuring air-quality management interventions, and serves as a communication tool on air-quality matters, to be easily understood and used by the public.

However, the report notes that a number of monitoring stations are not fully operational, and government does not provide any indication as to why there have been delays indicating poor risk communication and transparency. The two monitoring stations located in South Durban (namely, Settlers (Merebank) and Wentworth Reservoir stations) were not fully operational, and recordings were only available for sulphur dioxide and particulate matter (PM2.5 and PM10), but not for other pollutants such as nitric oxide, carbon monoxide, nitrogen dioxide, nitrogen oxides, benzene, toluene and xylene. Therefore, civil society did not have the necessary information to make informed decisions and act on potential pollutants in the air, harmful to their health and wellbeing, as provided for in the South African Constitution, or to hold government and industry accountable. Although the government has recognised shortcomings in its ability to maintain and monitor air quality, it has failed to adopt principles of ecological reflexivity to rethink activities surrounding monitoring, enforcement and compliance, which have potentially led to the continual pollution of the South Durban community, indicating a non-reflexive local governance institution.

However, the unreflective governance was also due to the lack of technical skills within the local government. Although skills were transferred to the new government during the transition in 1994, with many civil society leaders moving into different state structures, with the relationship between the state and civil society organisations characterised by a collaborative nature [19], several civil society informants noted that the lack of the eThekwini government skills had deteriorated since 2010. Some of the previously trained government officials had left the government sector to work in industry, or had moved to other sectors. This created a technical void within the government to effectively deal with industry and enforce regulations. Informant C noted that the government did not have the capacity and skills to effectively govern South Durban:

> 'Officials are not doing their job, we call them ... and they go to the industry, they can't read the information and the industry [then] tells them what to do and they [are] quite happy to believe that [as] they don't have the technical expertise ... There is also a cut back on qualified staff. They [local government] don't employ people who have the experience. Most of the air quality officers that were employed and trained in Denmark and Norway, they have also left and are either working for ENGEN or SAPREF or [now in] consultancy or they are in New Zealand, Australia or other parts of the world ... Since 2010, there has been a lack of credible data—the monitoring stations, fourteen of the most sophisticated and experienced stations has been allowed to decay and not work and not managed. So that is the problem ... the environment has not become an important issue by government at all costs and yet they want to make development decisions when the equipment are not operating .... It is a lack of leadership.' [Informant C, 3 July 2019].

The above statement suggests that there is not necessarily a lack of employed government officials, but rather not enough qualified, technically skilled staff. A review of the eThekwini Municipality Annual Report, 2017/2018 [70], shows that of the forty-one Environmental Protection posts, only one was vacant. A researcher in the field of occupational and environmental health, who has worked very closely with the South Durban community on health studies and risk exposure since 2002, stated that the lack of government technical skills, including their poor engagement and deliberation with civil society, influenced the relationship between civil society and the government. Civil society was not able to secure information on industrial pollution, due to poor communication and a lack of transparency from local government. As the informant noted:

> 'I definitely think that the skills set that they [local government] had around the MMP ... has been lost. There is an attempt to build a new technical skills base and to be honest, I don't know at this point, the levels of those individuals that are now there ... But what has happened is that because of our repeated attempts to get this sort of information that we want, there has not been that sort of forthcoming sort of relationship. We [civil society] have tended to lose that link that we have had [with government] over that period of time ... the air quality network ... [which] has been substantially compromised so the type of data we were getting in the 2000s when we were doing the health study is not as good as what we are seeing now ... It is local government who is supposed to maintain that [air quality stations] and run that and our sense is that there is a failing on government's side ... We have not been given clear reasons why these things are not working effectively' [Informant D, 20 June 2019].

6.2. Lack of Participatory Governance and Deliberation

Empirical analysis suggested that the government did not engage with reflexive governance principles by acknowledging the interpenetration of governance subjects and objects to engage more openly and directly with a variety of social actors to gather recursive feedback to understand social change. This governance approach was, therefore, divergent from the stated eThekwini Municipality city governance principles, as outlined in the eThekwini Municipality Annual Report, 2017/2018 [70]. The report highlights that some of the principles of good governance that the municipality demonstrates include 'accountability' as a 'fundamental requirement of good governance', and 'transparency', in that citizens should be able to follow and understand the decision-making process. Other principles outlined in the report include 'equity and inclusiveness', where community members 'feel their interests have been considered in the decision-making processes'. However, these principles have not mirrored the practices of local government. A closer analysis of the local government's 'participatory' governance principle suggests 'tokenistic' participatory approaches towards civil society. The local government's annual report highlights its participatory governance approach as follows, 'Anyone affected by or interested in a decision should have the opportunity to participate in the process of making that decision ... Community members may be provided with information, asked for their opinion, given the opportunity to make recommendations or, in some cases, be part of the actual decision-making process'. This approach suggests that when information is provided, and citizens are asked for their opinion, citizens may not be included from the onset of the strategy formulation and are only included in the final decision-making processes in select cases. This governance approach is tantamount to Arnstein's notion of top-down approach to governance by inviting citizens' opinions (namely, consultation), and thus offers no assurance that citizens' concerns will be taken into account [50]. It is also tantamount to 'placation', when citizens advise but local government retains the right to judge the legitimacy of the recommendations.

Poor reflective governance and a lack of government transparency regarding environmental risk information also influenced how civil society was able to meaningfully engage with local constituencies, government and industry to inform decisions. Informant D highlighted the lack of information from the government and how this compromised civil society actions regarding industrial risks:

> 'In the 2000s when the pollution control unit was setup and run by [Informant A] ... What is most important about that was their ability to make this information available on a short turnover basis. So, you could go to the website and you would know what was going on ... With my engagement with the organisations representing civil society I would say ... they don't have the information [from local government now] that they need to take the necessary action or to

engage with their stakeholders, the local communities. I think this is problematic. For example, for our birth cohort study—epidemiological study ... we needed emissions data, the emissions inventory. Now according to all our legislation and Access to Information Act, this data is supposed to be available. It took us a very, very long time, in fact probably more than five years. We had written to them [local government], we quoted the act and they agreed to release it to us, and we have repeatedly found there to be a reluctance to release this information. They said well this is confidential information' [Informant D, 20 June 2019].

This indicates a bureaucratic government approach, whereby modern approaches to reflexive governance since democracy spiralled into technocratic approaches. Local government has not enabled a transparent and inclusive governance by allowing civil society to express their needs and concerns. Despite civil society expressing their need for information, and as stipulated by regulations, the local government did not engage in ecological reflexivity by reflecting on their practices and changing their values in response to scrutiny and development solutions. Nor did the government spearhead democratic deliberations with stakeholders as a form of open communication to address diverse opinions. It was furthermore suggested by one informant that the government might not be concerned with facilitating deliberative democracy and engagement between industry and affected communities to address industrial risks. It was noted by a local government official that it was not necessarily the city department's responsibility to engage with local concerns and that there were local government structures in place to facilitate this engagement. However, this was not taking place due to politics, resulting in poor reflexive governance. According to a local government informant, namely, Informant E, there was a bias from government's political representatives regarding engagement with communities:

'It is not happening [local government bringing industry and communities together]... As local government, we have governance structures in place. We have our political reps, our ward councillors, who are the voice of the people and they should be doing that. They should be opening the channels of communication. They should be encouraging this. That is why SAPREF and the industries have included councillors in their forums so that they bring the voice of the people. That is tricky, councillors are supposed to have their community meetings. However, councillors will push the line of the people that support him or her because of politics. So, they will hear the voices of the people that support them, and the others may be seen as troublemakers or not so important.' [Informant E, 21 June 019].

Thus, the local government municipality has not engaged in effective participation and deliberation with civil society but has relied on local councillors to bring the concerns of the people to the municipal council. However, councillors may be inaccessible to people. Generally, Durban councillors have been accused of not communicating effectively with their constituents [73].

Councils may also be prone to corruption. For example, four eThekwini Municipality councilors were arrested in 2019 on charges of fraud and corruption [74]. The above government informant also acknowledges that local councillors may have vested interests. This indicates that the government recognises flaws in its strategy of communication and deliberation with civil society but has not been reflexive in rethinking practices and how to better engage with citizens in order to transform values and practices. A more open, transparent and inclusive governance is required to enable value reflexive governance to enable action and find specific solutions to problems.

Local government has not engaged with civil society to develop the instruments and procedures that would allow for a broader input into development processes, and to create feedback on development decisions made. Informant A emphasised the harsh approach by local government in not setting up consultative forums and, therefore, not applying the regulations of the country that enable consultations with civil society:

'Government can create spaces for people to come on board such as a forum, but it is not happening. Government is quite happy that it is not happening since they have less pressure on them. You have got the Constitution and all the regulations that say communities must be involved but they have not translated that into a modus-operandi to have a consultative forum ... if you collecting air quality information where is the report and for the information to go public ... Like the number of complaints, we have had for Mondi and SAPREF for the last year—what was the cause of that? How was it dealt with? Is it improving? The trends? So, you could be addressing social reporting' [Informant A, 19 June 2019].

Unfortunately, Informant D was less optimistic that the government would ensure proper consultation with civil society, due to the macro-economic development policies in the country, which drive business profits to the fore, at the expense of people and the environment. As the informant noted:

'So, this sort of information about knowledge transfer and making sure people are aware, we have to make sure it works. I think in a way I say that, but I know deep down it is not going to happen ... because it is such a divided society ... Those that are empowered believe that they are doing good for society with this trickle down effort and you will eventually benefit from this so let us go on and do it. I personally don't believe in that. It is breaking that barrier which is going to be a challenge. In a sense government is supposed to be the referee in this. They are supposed to say right, we are representing both stakeholders—we see the value in economic development, but we also see the impact on communities on civil society... But you don't see that happening because government then gets taken in because these things [industrial development] represent an increase tax base. So, the vested interest start[s] to shift ... ' [Informant D, 20 June 2019].

Some informants noted that this lack of transparency and participation from government was due to the close relationship between government and the industry for economic expansion and the industry making decisions for the government, especially with the latter not having the proper technical skills, as highlighted above. According to Informant C, the government has repeatedly made excuses for not releasing information because they work closely with industry, who control how government makes decisions:

'They [government] say under the pretext that it [information] is confidential. How can information that affects your health be confidential? They supposed to be revealing the information. During the apartheid era we did not get the information so why is it now during a democratic government they do not give you the information? What is critical is that there is a lack of political will. Previously, the national government and provincial and local government [after the transition to democracy] made sure that you got hold of the permits, the scheduled permits, you got the information. The air emission licenses are the most progressive permits that are ever done coming out of Norway. In Norway and Denmark, you can go to the government and they give you [the information] straight away about the facility. Here you can't get it and they won't allow it because the industry controls the government ... the folks that are working there [in local government] are too scare[d] to do their job for fear that the industry has got power and a lot of these politicians have got shares in these big companies ... that is the major problem' [Informant C, 3 July 2019].

It would seem, then, that during the transition to democracy, there was a new deposition for the government to enable a reflexive form of governance; however, this has deteriorated over the last decade due to a lack of political will and loss of technical skills. Informant F, who is the Air Quality/GIS & Youth Development Officer at the SDCEA, also noted the lack of government transparency, as well as the close relationship between the government and industry, which influenced the lack of proper governance:

'... the very same authorities, they are not giving us information. From 2010 to date, information that you used to get over the counter with no questions asked, even if now you can be a student, you want data, air quality monitoring data; they are going to ask you 101 questions. But before, they used to just give you data. So, I think there is a lot of secrecy ... You know the other biggest problem that we have seen is that, unfortunately authorities ... [are] paving their greener pastures to go to industries. That is the unfortunate part. They always want to look good on the side of industries. They don't want to be too harsh and too hard on industries.' [Informant F, 10 July 2017].

Due to the supposedly close relationship between government and industry, and lack of engagement from government towards civil society, some informants noted that major industries, such as the refineries, took advantage of this association and did not genuinely engage in participation with local civil society, since decisions on industrial development were already made (with government).

For example, the Engen Refinery Integrated Report, 2018 [76], noted that the industry strove to strengthen industry and government relationships by engaging on issues of mutual interest. Such engagements were noted to take place at Engen's senior leadership level, represented by the Chief Executive Officer. The industry was also part of the SAPIA 'to articulate and lobby the government to support the industry's positions'. Unfortunately, these engagements were separate from community engagement. In addition, the government did not enable a collective platform to bring all stakeholders together. A review of the SAPIA (2018) Annual Report contained a foreword by the government Minister of Energy, noting that national government had set a target to attract USD 100 billion of investment into the South African economy, and that the energy sector could contribute to a quarter of this target, as a minimum.

Reflexive governance engagement with civil society seems to be constrained by political power dynamics and macro-economic growth. This may imply that the government (and industry) are not sincerely concerned about deliberative democracy and collective governance with civil society and development processes. An academic human geographer from the University of KwaZulu-Natal, Informant G, researched social development in the area, and refers to the bureaucratic and tokenistic participation from the government (and industries) in the area as follows:

'The big issue [in South Durban] is lack of public participation. When participation took place it was top down and tokenism and almost telling you this is how it is going to be, rather than asking you what do you want? So, what is known as one of the most toxic zones in the world, the Merebank, Wentworth Zone will become more toxic ... ' [Informant F, 1 July 2019].

7. Discussion and Conclusions

This paper presented two main themes that emerged from grounded theory and coding of primary data. The first theme is the lack of governance skills and transparency and showed that poor governance, since 2010 and after the dismantling of the MPP, resulted in the local government not maintaining air-monitoring equipment to collect air-quality data to address potential industrial risks in South Durban. This has implications for the government not sharing air-quality information with civil society in order to make informed decisions about industrial development issues. This was a divergent approach from the previous MMP, launched after the dawn of democracy, which ran from 2000 to 2010, and resulted in improved air-quality monitoring and a reflexive government multi-stakeholder approach, with the government, community and industry working jointly. Since 2010, the lack of proper reflexive governance to maintain air-monitoring equipment has resulted in a lack of transparency and communication with civil society surrounding how the government collected the data. Although the government recognised shortcomings in its ability to maintain and monitor air quality, it failed to adopt principles of ecological reflexivity to rethink activities surrounding monitoring, enforcement and compliance, which potentially led to the continual pollution of the South Durban community, indicating a non-reflexive local governance institution. However, unreflective governance was also due to the lack of technical skills within the local government to effectively deal with industry and enforce regulations.

The second theme presented in the results surrounded the lack of participatory governance and deliberation. This showed that the government did not engage with reflexive governance principles by engaging transparently and directly with a variety of social actors to gather recursive feedback to understand social change. There were 'tokenistic' participatory approaches from the local government towards civil society, with the government retaining the right to judge the legitimacy of the recommendations put forward by civil society rather than engaging with them on an equal footing. Thus, poor reflective governance and a lack of government transparency on environmental risk information influenced how civil society was able to engage with local constituencies, government and industry to inform decisions. Modern approaches to reflexive governance since the establishment of democracy have spiraled into technocratic approaches. Local government did not view it as their responsibility to engage with local concerns, but the duty of inefficient local government structures to facilitate this engagement. Although government recognised flaws in its strategy of communication and deliberation with civil society, it was not reflexive in rethinking practices and how it better engaged with citizens. Some informants noted that a lack of transparency and participation from government was due to the close relationship between government and industry for economic expansions and due to industry dominating development decisions. Major industries took advantage of the close relationship with government and did not genuinely engage in participation with local civil society since industrial development decisions were already made with government. Thus, reflexive governance to engage with civil society was found to be constrained by political power dynamics and macro-economic growth.

Overall, this paper has shown that there are a number of constraints involved in enabling reflexive governance, particularly in terms of development and middle-upper-income countries such as South Africa. The country's democratic transition has enabled the supportive governance policy frameworks and the overall guiding principle of reflexive governance values. The newly elected government has instituted a democratic constitution and developed various regulations to strengthen democracy and accountability, for example by ensuring the participation of civil society in many political issues. Despite this positive development, there has been a very restricted implementation of value reflexive policies, not least at the local government level. The analysis found three crucial reasons for this: lack of sufficient skills at the local governmental level, the local government's reliance on technical experts from the industry, and the lack of platforms where government, industry and civil society can collectively deliberate on industrial risk issues.

Reflexive governance application in the South Durban political contexts had implications for its workings and efficiency. Civil society expressed uncertain relations with local government and questioned the democratic legitimacy of its value reflexive governance approaches. Poor reflexive governance was due to a lack of local government skills in maintaining air-monitoring equipment, and not addressing any potential industrial risks in South Durban. This, in turn, had implications for the government sharing air-quality information with civil society to make informed decisions about industrial development issues, with implications on participation and transparency. Although the government recognised shortcomings in its ability to maintain and monitor air quality, it failed to engage in ecological reflexivity and rethink activities surrounding monitoring, enforcement and compliance, indicating a non-reflexive local governance approach.

The government, as an enforcer of the law, relied on industry technical expertise, and thus spiralled into biased and technocratic approaches removed from the public. Some of the previously trained government officials left the government sector to work in industry or moved to other sectors. This created a technical void within the government to effectively deal with industry and enforce regulations. The eThekwini government skills have rapidly deteriorated since 2010, after the conclusion of the MPP, and thus weakened reflexive governance. These shortcomings have seriously hindered the way the government operates and engages with both civil society and industry and, hence, the implementation of reflective governance, the development of value reflective governance principles for engagement with civil society and industry, and the spearheading of ecological reflexivity to improve its operations and practices. These have had implications for upholding value reflexive governance principles outlined in the South African Constitution, including other environmental and governance policies and regulations.

The further failure of a value-reflexive governance and ecological reflexivity approach was that the local government did not engage with reflexive governance principles by creating platforms where the civil society, government and industry could openly communicate and make joint decisions, nor was the government concerned with gathering recursive feedback to understand social change. Therefore, this governance approach was divergent from the stated eThekwini Municipality city governance principles. This lack of engagement compromise civil society's ability to meaningfully engage with local constituencies, government and industry to inform decisions. The municipality also relied on local councils to bring the concerns of the people to the municipal council. However, councillors were inaccessible to people or prone to corruption. A more open, transparent and inclusive governance approach is required to enable a value-reflexive governance approach to enable action and find specific solutions to problems. This will require the government to enable the appropriate deliberative platforms to jointly engage with civil society and industry to obtain a consensus surrounding the relevant instruments and procedures, which would allow for a broader input into development processes, and feedback on the development decisions made. At the present moment, this is absent and has enabled tokenistic participatory and communicative approaches towards civil society when presented.

The emerging democratic structure of South Africa creates opportunities to better handle industrial environmental risks. As shown in this paper, a reflexive governance approach has the potential to address these risks in a democratically sound and environmental relevant way. A prerequisite, however, is that the government gives priority and allocates resources to this approach. As this case study indicated, certain conditions need to be met. First of all, it requires that the government does not prioritise neoliberal profit driven by industrial motives at the expense of social and environmental concerns. Due to a governmental, neoliberal development paradigm, old modes of governance are still present and exercised within the framework of reflexive governance policy and principles. Secondly, this requires that the local government develop a sufficient technical capacity so that it may also be able to make informed development decisions, without relying on industrial expertise. Thirdly, it requires that the local government, as enforcer of the law, maintains a neutral approach by not being one-sidedly influenced by industrial interests, and by providing a platform for civil society to inform decisions. This will require that those employed within local government have the required skills and expertise and are able to make independent decisions surrounding industrial development.

Combining the value-reflexive governance and ecological reflexivity will foster the formulation of values and the development of the instruments and procedures that can better incorporate diverse opinions into development decisions. Thus, the local government must engage in ecological reflexivity by rethinking its practices and transforming how it currently enables reflexive governance. Such an approach will enable a stronger reflexive governance approach. However, the challenge is not only to build relevant local governance structures, but also to develop new interfaces between the government and society, and find a balance of power between stakeholders. Generally, this case from South Africa has highlighted that reflexive governance practitioners need to be cognisant of its applicability across diverse geographic settings and beyond western notions of reflexive governance.

Author Contributions: Conceptualization, L.L. and R.L.; methodology, L.L.; validation, R.L.; formal analysis, L.L. and R.L.; investigation, L.L.; writing—original draft preparation, L.L.; writing—review and editing, L.L. and R.L.; supervision, R.L.; project administration, L.L. All authors have read and agreed to the published version of the manuscript.

Funding: This research received no external funding.

Institutional Review Board Statement: The study was conducted according to the guidelines of the Declaration of Helsinki, and approved by the College Ethics Committee) of the University of South Africa, College of Agriculture and Environmental Sciences.

Informed Consent Statement: Informed consent was obtained from all subjects involved in the study.

Data Availability Statement: Not applicable.

Conflicts of Interest: The authors declare no conflict of interest.

References

1. Newig, J.; Voß, J.; Monstadt, J. Editorial: Governance for Sustainable Development in the Face of Ambivalence, Uncertainty and Distributed Power: An Introduction. *J. Environ. Policy Plan.* **2007**, *9*, 185–192. [CrossRef]
2. Bäckstrand, K.; Kuyper, J.; Linnér, B.; Lövbrand, E. Non-state actors in global climate governance: From Copenhagen to Paris and beyond. *Environ. Politics* **2017**, *26*, 56–579. [CrossRef]
3. Dryzek, J.; Pickering, J. Deliberation as a catalyst for reflexive environmental governance. In *Working Paper Series 3*; Centre for Deliberative Democracy and Global Governance: Canberra, Australia, 2016.
4. Voß, J.-P.; Bornemann, B. The Politics of Reflexive Governance: Challenges for Designing Adaptive Management and Transition Management. *Ecol. Soc.* **2011**, *16*. Available online: http://www.ecologyandsociety.org/vol16/iss2/art9/ (accessed on 2 June 2020). [CrossRef]
5. Gunningham, N. Regulatory reform and reflexive regulation: Beyond command and control. In *Reflexive Governance for Global Public Goods, Politics, Science and the Environment*; Brousseau, E., Dedeurwaerdere, T., Siebenhüner, B., Eds.; MIT Press: Cambridge, MA, USA, 2012; pp. 85–104.
6. Swyngedouw, E.; Kaika, M. Urban political ecology. Great promises, deadlock ... and new beginnings? *Doc. d'Anàlisi Geogràfica* **2014**, *60*, 459–481.
7. Dagkas, A.; Tsoukala, K. Civil society, identity, space and power in the neoliberal age in Latin America and Greece. *Synergies* **2011**, *3*, 105–113.
8. Rosenau, J.N. Governance, Order and Change in World Politics. In *Governance without Government: Order and Change in World Politics*; Rosenau, J.N., Czempiel, E.-O., Eds.; Cambridge University Press: Cambridge, UK, 1992; Chapter 1; pp. 1–29.
9. Luna, V. From Neoliberalism to Possible Alternatives. *Inf. Econ.* **2015**, *395*, 35–49. [CrossRef]
10. Santos, T. Confronting governance challenges of the resource nexus through reflexivity: A cross-case comparison of biofuels policies in Germany and Brazil. *Energy Res. Soc. Sci.* **2020**, *65*, 101464. [CrossRef]
11. Pickering, J. Ecological reflexivity: Characterising an elusive virtue for governance in the Anthropocene. *Environ. Politics* **2019**, *28*, 1145–1166. [CrossRef]
12. Weiland, S. Reflexive governance: A way forward for coordinated natural resource policy? In *Environmental Governance: The Challenge of Legitimacy and Effectiveness*; Hogl, K., Kvarda, E., Nordbeck, R., Pregernig, M., Eds.; Edward Elgar Publishing: Cheltenham, UK, 2012.
13. McNutt, K.; Rayner, J. Valuing Metaphor: A constructive account of reflexive governance in policy networks. In Proceedings of the 5th International Conference on Interpretative Policy Analysis, Grenoble, France, 23–25 June 2010.
14. Meadowcroft, J.; Steurer, R. Assessment practices in the policy and politics cycles: A contribution to reflexive governance for sustainable development. *J. Environ. Policy Plan.* **2018**, *20*, 734–775. [CrossRef]
15. Boström, M.; Lidskog, R.; Uggla, Y. A reflexive look at reflexivity in environmental sociology. *Environ. Sociol.* **2017**, *3*, 6–16. [CrossRef]
16. Guay, T. Local Government and Global Politics: The Implications of Massachusetts' "Burma Law". *Political Sci. Q.* **2000**, *115*, 353–376. [CrossRef]
17. Leonard, L. Reconsidering the 'risk society theory' in the South. *S. Afr. Rev. Sociol.* **2014**, *45*, 74–93.
18. Maguranyanga, B. South African Environmental Justice Struggles against 'Toxic' Petrochemical Industries in South Durban. Available online: http://www.umich.edu/~{}snre492/brain.html (accessed on 17 March 2011).
19. Leonard, L.; Pelling, M. Civil Society Response to Industrial Contamination of Groundwater in Durban, South Africa. *Environ. Urban.* **2010**, *22*, 579–595. [CrossRef]
20. South African Constitution. Republic of South Africa, Act 108 of 1996. Available online: https://www.gov.za/sites/default/files/images/a108-96.pdf (accessed on 2 June 2020).
21. National Waste Management Strategy. Department of Environmental Affairs and Tourism, Republic of South Africa. Available online: http://www.polity.org.za/pol/acts/ (accessed on 2 June 2020).

22. National Environmental Management Act, Act 107, Department of Environmental Affairs and Tourism, Republic of South Africa. Available online: https://www.environment.co.za/documents/legislation/NEMA-National-Environmental-Management-Act-107-1998-G-19519.pdf (accessed on 2 June 2020).
23. Mathee, A. Environment and Health in South Africa. *J. Public Health Policy* **2011**, *32*, 37–43. [CrossRef] [PubMed]
24. Malherbe, S.; Segal, N. Corporate Governance in South Africa. In Proceedings of the Policy Dialogue Meeting on Corporate Governance in Developing Countries and Emerging Economies, OECD Development Centre and the European Bank for Reconstruction and Development, OECD headquarters, Muldersdrift, South Africa, 23–24 April 2001.
25. Ballard, R.; Habib, A.; Valodia, I.; Zuern, E. Globalization, marginalization and contemporary social movements in South Africa. *J. Afr. Aff.* **2005**, *104*, 615–634. [CrossRef]
26. Barchiesi, F. *Classes, Multitudes and the Politics of Community Movements in Post-Apartheid South Africa*; Centre for Civil Society Research: Berlin, Germany, 2004.
27. Makino, K. Institutional conditions for social movements to engage in formal politics. In *Protest and Social Movements in the Developing World*; Shigetomi, S., Kumiko, M., Eds.; Edward Elgar Publishing: Cheltenham, UK, 2009.
28. Leonard, L. Civil Society Reflexiveness in an Industrial Risk Society. Ph.D. Thesis, University of London (Kings College), London, UK, 2009.
29. Fig, D. Manufacturing amnesia. *Int. Aff.* **2005**, *81*, 599–617. [CrossRef]
30. Friedman, S. Participatory Governance and Citizen Action in Post-Apartheid South Africa. In *Discussion Paper Series No*; International Institute for Labour Studies: Geneva, Switzerland, 2006.
31. Beck, U. *Risk Society*; SAGE Publications: Thousand Oaks, CA, USA, 1992.
32. Beck, U.; Giddens, A.; Lash, S. *Reflexive Modernization*; Polity Press: Cambridge, UK, 1994.
33. Rosa, E.; Renn, O.; McCright, A. *The Risk Society Revisited: Social Theory and Governance*; Temple University Press: Philadelphia, PA, USA, 2014.
34. Beck, U. *The Metamorphosis of the World*; Polity Press: Cambridge, UK, 2016.
35. Giddens, A. *The Consequences of Modernity*; Polity Press: Cambridge, UK, 1990.
36. Visvizi, A. Safety, risk, governance and the Eurozone crisis: Rethinking the conceptual merits of 'global safety governance'. In *Essays on Global Safety Governance: Challenges and Solutions*; Kłosińska-Dąbrowska, P., Ed.; University of Warsaw Publishing Programme: Warsaw, Poland, 2015.
37. Kunseler, E. Revealing a paradox in scientific advice to governments. *Palgrave Commun.* **2016**, *2*, 1–9. [CrossRef]
38. Kooiman, J.; Bavinck, M.; Chuenpagdee, R.; Mahon, R.; Pullin, R. Interactive Governance and Governability: An Introduction. *J. Transdiscipl. Environ. Stud.* **2008**, *7*, 1–11.
39. Weston, D.; Goga, S. *Natural Resource Governance Systems in South Africa. Water Research Commission*; Department of Water and Sanitation: Pretoria, South Africa, 2016. Available online: http://www.wrc.org.za/wp-content/uploads/mdocs/WIN1_IWRM.pdf (accessed on 2 June 2020).
40. Marissing, V.E. Citizen participation in the Netherlands Motives to involve citizens in planning processes. In Proceedings of the ENHR Conference 'Housing: New Challenges and Innovations in Tomorrow's Cities', Reykjavik, Iceland, 29 June–3 July 2005.
41. Rogers, P.; Hall, A.W. *Effective Water Governance*; Global Water Partnership Technical Committee, TEC Background Papers; Elanders: Mölndal, Sweden, 2003; Volume 7.
42. Tembo, F. *Participation, Negotiation, and Poverty: Encountering the Power of Images*; Ashgate Publishing Company: Farnham, UK, 2003.
43. Graham, D.; Amos, B.; Plumptre, T. Principles for Good Governance in the 21st Century. In Proceedings of the Fifth World Parks Congress, Durban, South Africa, 8–17 September 2003.
44. Wesselink, A.; Paavola, J.; Fritsch, O.; Renn, O. Rationales for public participation in environmental policy and governance: Practitioners' perspectives. *Environ. Plan.* **2011**, *43*, 2688–2704. [CrossRef]
45. Zeijl-Rozema, A.; Cörvers, R.; Kemp, R. Governance for sustainable development: A framework. In Proceedings of the Earth System Governance: Theories and Strategies for Sustainability, Amsterdam, The Netherlands, 24–26 May 2007.
46. Blowers, A. Environmental Policy: Ecological Modernisation or the Risk Society? *Urban Stud.* **1997**, *34*, 845–871. [CrossRef]
47. Leonard, L. Governance, participation and mining development. *Politikon* **2017**, *44*, 327–345. [CrossRef]
48. Orr, R. Governing When Chaos Rules: Enhancing governance and participation. *Wash. Q.* **2002**, *25*, 139–152. [CrossRef]
49. Ishii, R.; Farhad, H.; Rees, C. Participation in Decentralized Local Governance: Two Contrasting Cases from the Philippines. *Public Organ. Rev.* **2007**, *7*, 359–373. [CrossRef]
50. Arnstein, R.S. A Ladder of Citizen Participation. *J. Am. Inst. Plan.* **1969**, *35*, 216–224. [CrossRef]
51. Stirling, A. Precaution, foresight, sustainability: Reflection and reflexivity in the governance of science and technology. In *Reflexive Governance for Sustainable Development*; Voß, J.P., Bauknecht, D., Kemp, R., Eds.; Edward Elgar: Cheltenham, UK, 2006; pp. 225–272.
52. Walker, G.; Shove, E. Ambivalence, Sustainability and the Governance of Socio-Technical Transitions. *J. Environ. Policy Plan.* **2007**, *9*, 213–225. [CrossRef]
53. Sonnino, R.; Torres, C.; Schneider, S. Reflexive governance for food security: The example of school feeding in Brazil. *J. Rural Stud.* **2014**, *36*, 1–12. [CrossRef]
54. Voß, J.-P.; Kemp, R. Sustainability and reflexive governance: Introduction. In *Reflexive Governance for Sustainable Development*; Voß, J.-P., Bauknecht, D., Kemp, R., Eds.; Edward Elgar: Cheltenham, UK, 2006.

55. Meisch, S.; Beck, R.; Potthast, T. Towards a value reflexive governance of water. In *Climate Change and Sustainable Development: Ethical Perspectives on Land Use and Water Production*; Potthaust, T., Meisch, S., Eds.; Wageningen Academic Publishers: Wageningen, The Netherlands, 2012.
56. Smith, A.; Stirling, A. Moving Outside or Inside? Objectification and Reflexivity in the Governance of Socio-Technical Systems. *J. Environ. Policy Plan.* **2017**, *9*, 351–373. [CrossRef]
57. Kaltenborn, P.; Qvenild, M.; Nellemann, C. Local governance of national parks: The perception of tourism operators in Dovre-Sunndalsfjella National Park, Norway. *Nor. Geogr. Tidsskr. Nor. J. Geogr.* **2011**, *65*, 83–92. [CrossRef]
58. Wiley, D.; Root, C.; Peek, B. Contesting the urban industrial environment in South Durban in a period of democratisation and globalisation. In *Durban Vortex*; Freund, B., Padayachee, V., Eds.; University of Natal Press: Pietermaritzburg, South Africa, 2002; pp. 223–254.
59. Scott, D. Creative destruction: Early modernist planning in the South Durban industrial zone, South Africa. *J. S. Afr. Stud.* **2003**, *29*, 235–259. [CrossRef]
60. Leonard, L.; Lidskog, R. Industrial Scientific Expertise and Civil Society Engagement: Reflexive Scientisation in the South Durban Industrial Basin, South Africa. Available online: https://www.tandfonline.com/doi/abs/10.1080/13669877.2020.1805638 (accessed on 1 September 2020).
61. Reid, K.; D'Sa, D. The double edged sword: Advocacy and lobbying in the environmental sector. *Crit. Dialogue* **2005**, *2*, 1–39.
62. Dwyer, P. Wentworth and the South Durban Industrial Basin. *Alternatives*, 20 January 2004. Available online: http://www.alternatives.ca/article1240.html (accessed on 20 October 2008).
63. D'Sa, D.; Bond, P. Odious Brics Loan for Durban Port Project will Not Go Unopposed. *Business Day*, 6 June 2018. Available online: https://www.businesslive.co.za/bd/opinion/2018-06-06-odious-brics-loan-for-durban-port-project-will-not-go-unopposed/ (accessed on 1 September 2020).
64. Kistnasamy, E.J.; Robins, T.G.; Naidoo, R.; Batterman, S.; Mentz, G.B.; Jack, C.; Irusen, E. The relationship between asthma and ambient air pollutants among primary school students in Durban, South Africa. *Int. J. Environ. Health* **2008**, *2*, 365–385. [CrossRef]
65. South Durban Community Environmental Alliance. 2017. Available online: https://sdcea.co.za/2017/03/31/a-brief-compilation-of-major-pollution-incidents-in-the-south-durban-basin-from-2000-2016/#post/0 (accessed on 9 September 2019).
66. Leonard, L. Oil Refinery Blasts Is One More Reason South Africa Should Take Industrial Risks Seriously. *The Conversation*, 11 December 2020. Available online: https://theconversation.com/oil-refinery-blast-is-one-more-reason-south-africa-should-take-industrial-risks-seriously-151779 (accessed on 1 September 2020).
67. Pillay, N. Durban South Basin a Hotspot for High Air Pollution. *SABC News*, 7 May 2018. Available online: https://www.sabcnews.com/sabcnews/durban-south-basin-a-hotspot-for-high-air-pollution/ (accessed on 1 September 2020).
68. Leonard, L.; Nunkoo, R. Examining 'toxic tourism' as a new form of alternative urban tourism and for environmental justice: The case of the South Durban Industrial Basin, South Africa. In *Urban Tourism in Sub-Saharan Africa: Risk and Resilience*; Leonard, L., Musvengence, R., Saikwah, P., Eds.; Taylor and Francis: Abingdon-on-Thames, UK, 2021; pp. 17–30.
69. South Durban Basin MPP Case Study Report. A Governance Information Publication, Department of Environmental Affairs and Tourism. 2007, Series C. Available online: http://www.airqualitylekgotla.co.za/assets/south_durban_basin_multi-point_plan_case_study_report.pdf (accessed on 30 April 2020).
70. eThekwini Municipality Annual Report. 2017. Available online: http://www.durban.gov.za/Resource_Centre/reports/Reports/PM%20Reports/Forms/AllItems.aspx?RootFolder=%2fResource%5fCentre%2freports%2fReports%2fPM%20Reports%2fAnnual%20Report%202017%2d2018&FolderCTID=0x012000E24596D52A5CDD4B9E95F1C77CE7CC64 (accessed on 1 May 2020).
71. Mngadi, S. City of eThekwini in Air Quality Monitors Dispute. *Independent Online*, 16 June 2019. Available online: https://www.iol.co.za/news/south-africa/kwazulu-natal/city-of-ethekwini-in-air-quality-monitors-dispute-26412057 (accessed on 1 September 2020).
72. National Air Quality Indicator—Monthly Data Report for the KwaZulu Natal Province. Available online: https://saaqis.environment.gov.za/ (accessed on 29 April 2020).
73. Mail and Guardian. Ward Councillors Are Key Players. 12 October 2018. Available online: https://mg.co.za/article/2018-10-12-00-ward-councillors-are-key-players/ (accessed on 2 May 2020).
74. Africa News Agency. Councillors Linked to R208m DSW Tender Scandal Released on R5000 Bail. 11 December 2019. Available online: https://www.iol.co.za/news/south-africa/kwazulu-natal/councillors-linked-to-r208m-dsw-tender-scandal-released-on-r5000-bail-39082898 (accessed on 2 May 2020).
75. South African Petroleum Industry Association. Annual Report. Available online: https://www.sapia.org.za/Portals/0/Annual-Reports/SAPIA_AR%202018_FA.pdf (accessed on 10 May 2020).
76. Engen Refinery Integrated Report. Available online: http://www.engen.co.za/Media/Default/PDF/018-2018%20Integrity%20Report%20Final%2024%20July.pdf (accessed on 10 May 2020).
77. Creswell, J. *Research Design: Qualitative, Quantitative and Mixed Methods Approaches*; SAGE Publications: Thousand Oaks, CA, USA, 2014.

 *sustainability*

Article

Public Policy for Social Innovations and Social Enterprise—What's the Problem Represented to Be?

Jörgen Johansson [1,*] and Jonas Gabrielsson [2]

1. School of Public Administration, University of Gothenburg, SE-405 30 Gothenburg, Sweden
2. School of Business, Innovation and Sustainability, Halmstad University, SE-301 18 Halmstad, Sweden; jonas.gabrielsson@hh.se
* Correspondence: jorgen.johansson@spa.gu.se

Abstract: Social innovations and social enterprise have been seen as innovative measures to achieve sustainable development. Drawing on an evaluation of a development project on creating social enterprises in Sweden, this article analyzes social innovations as a policy area. The policy area is often described as loaded with ideological contradictions. The aim of the article is to explore underlying premises and discourses in policy implementation aimed at creating social innovations in a comparison between two ideal types on social sustainability—(1) an individual activation strategy (responsibilization of the individual) and (2) a societal equilibrium strategy (balancing social values). The research question is inspired by Carol Bacchi's policy theory and asks what is the problem represented to be? The analysis is carried out at the micro-level as a context-sensitive approach to explore articulations made among actors creating the policy and entrepreneurs participating in a locally organized project. The article contribute with a better understanding of how societal problems and their solutions are discursively determined, with implications for policy makers and project managers active in this policy area. The analysis and findings indicate a significant policy shift during the implementation process. Initially, the policy idea consisted of well-considered ambitions to create a long-term sustainable development. During the implementation of the project, the problem's representation changes gradually in the direction towards individual activation. This transition is driven by pragmatic difficulties of defining the policy area, problems of separating means from ends, and the need to make decisions based on a limited range of information. We conclude by emphasizing the need for reflection on how the social dimension is defined when implementing social innovation strategies. Furthermore, there is a lack of studies of how this policy area can be linked to policies for social sustainability.

Keywords: social innovation; social enterprise; policy analysis; problem representation; individual activation; social sustainability

Citation: Johansson, J.; Gabrielsson, J. Public Policy for Social Innovations and Social Enterprise—What's the Problem Represented to Be? *Sustainability* **2021**, *13*, 7972. https://doi.org/10.3390/su13147972

Academic Editors: Ingemar Elander and Andrea Pérez

Received: 26 May 2021
Accepted: 13 July 2021
Published: 16 July 2021

Publisher's Note: MDPI stays neutral with regard to jurisdictional claims in published maps and institutional affiliations.

Copyright: © 2021 by the authors. Licensee MDPI, Basel, Switzerland. This article is an open access article distributed under the terms and conditions of the Creative Commons Attribution (CC BY) license (https://creativecommons.org/licenses/by/4.0/).

1. Introduction

During the last two decades, there have been monumental hopes for social innovation to achieve sustainable development. In 2009, President Obama launched the Social Innovation Fund to support initiatives in doing business differently by promoting community leadership and investments in innovative community solutions. In Japan, social innovation has been a part of the rebuilding efforts following the 2011 nuclear disaster, which left massive destruction on its assets such as the physical and sociopolitical environment. In the United Kingdom, the Office of Civil Society is designed to enrich lives, drive growth, and promote Britain to the world by working in partnerships with civil society, private businesses, and the state. Recently, social innovation has also been included in the EU 2020 strategy for smart, sustainable, and inclusive growth. Public policy efforts in enhancing social innovations and social enterprises have thus been seen as complementary measures to help solve many of the contemporary problems in a situation where public budgets are under pressure, and public policies are suffering from sectorization and fragmentation [1–8].

Drawing on an evaluation of a regional development project on creating new social enterprises in Sweden, this article analyzes social innovations and social enterprises as an emerging policy area. Policies in this field are often described as both dynamic and complex [9–12]. Firstly, scholars and practitioners in the policy field disagree on definitions [13]. Secondly, and despite the conceptual unanimity, social innovations and social enterprises are policy concepts associated with essential hybridity. They exist in a territory in between the for-profit and nonprofit sectors, and they often combine the logics of the spheres of the state, the market (for-profit), and the civil society (including the community and organizations in the third, non-profit sector) [14,15]. This hybridity seems to create ambivalence among policymakers and participating organizations. For example, the introduction of projects enhancing social enterprises is often met with skepticism among stakeholders in the business sector. On the other hand, actors in civil society are constantly playing the role of the energetic proponent. Still, they often experience disappointment in the slow progress of change or, in the worst case, lost opportunities to reach long-term effects. In the middle stands state authorities and municipal actors that try to mediate between, on one side, demands on a market-oriented and commercial approach and, on the other side, claims for acting following social and human values. In other words, the policy area is loaded with ideological conflicts and contradictory ambitions [16–20].

In this article, we will focus on especially one kind of contradiction embedded in the policy field, namely between two opposing perspectives on social sustainability—on one side, *an individual activation strategy*, and, on the other side, *a societal equilibrium strategy*. The two strategies will be compared by an ideal-typical comparative analysis that explicates different ways of articulating the policy problem. Firstly, in the activation strategy, social sustainability contains policies that aim at implanting in individuals' the interest to promote their employability, life-long learning, and attaining the "right" attitudes, e.g., flexibility, career aspirations, entrepreneurial mindsets [21]. Secondly, the societal equilibrium strategy is based on policies that aim to create an equilibrium between often contradictory social values, e.g., justice, human development, and security. Strong social sustainability correlates with a high degree of equilibrium between these contradictory values [22]. In Section 3, we will give a more extensive description of the theories and methods of the comparative analysis.

Our inquiry is guided by the following research question: What is the problem represented to be in policies enhancing social innovations and social enterprises and what is silenced in this representation? To address our research question, we will analyze three sub-questions:

- In what ways are the societal reality characterized in the representation?
- What normative assumptions underpin the representation?
- How has the representation of the problem been transformed by policy proposals?

Against this, the overall purpose of the article is to explore central issues in policies aimed at creating social innovations and social enterprises in a comparison between two ideal types representing different approaches to the policy problem. In that way, the comparative analysis will include considerations on issues of power relations among participating actors and producing legitimacy towards beneficiaries, e.g., presumptive entrepreneurs, non-profit organizations, trade unions, and organizations representing economic and social interests.

The comparative analysis will be carried out at the micro-level as we are interested in how local actors "live" the policy area. Our context-sensitive approach enables us to identify and explore articulations made among actors creating the policy and social entrepreneurs participating in locally organized development projects. Moreover, our research question focuses on the underlying premises, assumptions, and discourses in policy implementation [23]. We employ a research design that involves multiple sources of data, such as documents preparing the policy, interviews with project managers and social entrepreneurs, as well as participatory observations. By analyzing how the involved actors problematize and motivate the need for creating social innovations and supporting social enterprises, we can collect rich 'bottom-up' articulations subject to ideal-typical comparative analysis.

Our findings suggest that the original intentions articulated at the regional level and expressed in the project application become something partly different as processes and activities become implemented at lower levels in the implementation chain. Our multi-level analysis of the process shows how idealistic ambitions aimed at creating a long-term sustainable development of society are filtered through pragmatic difficulties of defining values and objectives, separating means from ends, and making decisions based on a limited range of information and analysis. Moreover, our findings illustrate how locally organized social enterprising efforts, championed by entrepreneurs in the project, struggle with managing contradictory values, fragmented organizational structures, and scarce resources. In this respect, our findings contribute to literature and research on social sustainability by providing theoretical and empirical insights of issues and challenges involved when creating and implementing social innovations as a specific policy area in a regional setting

The article is organized in the following way. In Section 2 follows a background describing some essential characteristics of the policy area on social innovations and social enterprises. In the following Section 3, we will present the theoretical and methodological framework. In Section 4, the findings are presented, and the research questions are answered. In Section 5, we finally conclude with a discussion of the main findings.

2. Social Innovation and Social Enterprise—Concepts and Policy Area

The concepts of social innovation and social enterprises are contested in research as well as in political life [24–27]. Social innovation is often defined as new ideas, products, services, and methods that meet social challenges. These can be climate issues, integration, unemployment, an aging population, and social exclusion [27], or a more developed definition such as:

> a novel solution to a social problem that is more effective, efficient, sustainable, or just than existing solutions and for which the value created accrues primarily to society as a whole rather than private individuals. A social innovation can be a product, production process, or technology (much like innovation in general), but it can also be a principle, an idea, a piece of legislation, a social movement, an intervention, or some combination of them (p. 36, [28]).

In this definition, it is explicitly underlined that social innovations accrue primarily to societal issues rather than on individuals and with primacy given to social over economic value creation. Additionally, the understanding of social enterprise remains debated amongst scholars as well as in practice [14,29,30]. It is widely diffused that social enterprise encompass organizing efforts with a central mission to have a transforming impact or to create positive social change [31]. However, definitions continue to range from broader to more narrow approaches. Broadly defined, social enterprise refers to innovative activity with a social objective in either the for-profit sector, the non-profit sector, or both. Narrowly, it refers to simply applying market-based skills and commercial activities in the non-profit sector to create social value and addressing social or environmental needs.

The main conceptual problem associated with social innovation and social enterprise as a policy area is finding a basis for determining what is social and what is not. The social denotes very different things; social motivations or intentions, the social as based on ideals in a community, and processes in society that create social value. The social is also commonly equated with the societal problems or challenges it tries to solve [32,33].

Another kind of conceptual orientation asserts that social enterprise has its specificity in that it simultaneously stresses both the process and results in enterprising efforts [14]. Traditional for-profit enterprise is mainly fixated on the result, but social enterprise encompasses values such as participation, solidarity, trust, and learning as important as the results. Hence, the intrinsic value of the processes, not only its output or results, is often emphasized in social enterprise development aimed at social innovations [31,34]. An important point of departure in this article is to examine how social innovations and social enterprising efforts differ from more traditional forms of for-profit-based innovation and entrepreneurship. How do the actors motivate the social in their venturing efforts, and what problems do they intend to solve?

During the last two decades, the concepts of social innovation and social enterprise have gained significant recognition as a new policy area within the broader spectrum of industrial development policies in most countries worldwide. Some scholars conclude that this policy area represents a new paradigm that transcends the traditional boundaries between state, market, and civil society [35,36]. The paradigm has its roots in at least two kinds of movements. Firstly, it is connected to the cooperative movement and ideas referred to as the social economy [37]. In this context, the term social enterprises were created and defined as enterprises built on three dimensions [38]:

1. Economic dimensions (market orientation, risk taking);
2. Social dimensions (utilization of resources to communities and for welfare provision);
3. Participative dimensions (involvement of users, room for deliberation, transparency).

In this policy context, it is also worth highlighting the EU Social Business Initiative presented in 2011 by the European Commission. The initiative established an EU-level action plan with concrete measures to develop a favorable environment for social enterprises. This initiative is yet another expression of a policy area that is at an emerging stage. In a research report, Defourny & Nyssens note that initiatives in this policy field are on the rise, and they conclude:

> The debate is now on both the public and the private agenda. Indeed, both the public sector and the private sector, each in its own way, are discovering or rediscovering new opportunities to promote, simultaneously, entrepreneurial spirit and the pursuit of the public good (p. 32) [39]. (See also [40])

Secondly, it is connected to an international movement consisting of influential NGOs, foundations, networks, etc., promoting ideas to integrate social innovations in various efforts to enhance social values in economic and societal development. Among the most influential foundations is the Ashoka Foundation, the Schwab Foundation and the Skoll Foundation, which have made large investments in social innovation, often labeled as 'venture philanthropy' [41,42].

These ideas have diffused globally, and in many countries, governments have institutionalized social innovation and social enterprise as a specific policy area. In the Nordic countries, the introduction of policies promoting social innovations is still regarded as embryonic. In a research report presented for the Nordic Council of Ministers, four kinds of shared characteristics are identified in public policies promoting social innovation and social enterprises in the Nordic Countries [43]:

1. The welfare states are an innovative and active partner to develop this policy area.
2. A basic policy idea is that social enterprises are built on co-operation between the public, private, and civil society.
3. The policy area includes much more than activities tied to work integration
4. The policy area function as arenas for citizens' participation, learning and provision of welfare services.

If we look at Sweden, policies enhancing social innovations and social enterprises do not have any specific legal framework or comprehensive documentation (strategies, organizational structures, resources) that illustrate the scope, orientation, and development of social innovations and social enterprises. Policies enhancing this area extend across several policy sectors concerning, e.g., regional growth, industry and trade, labour market, academic research, politics for civil society (see [44,45]). In 2017, the Swedish government launched a strategy on social enterprises—*A sustainable society through social enterprise and social innovation* [46]. This strategy has one overall goal and five specific areas that will be cornerstones in the future development of social enterprises in Sweden. The overall goal is to strengthen the development of social enterprises to better take part in solving challenges in society and contribute to efforts in the public sector to recognizes and make use of social enterprises as valuable actors in a sustainable society. In the strategy, the government identifies five kinds of specific policy measures that aim to coordinate and strengthen a wide range of components that enhance the development of social innovations to:

1. create needs and demands in the public sector to support social innovations,
2. improve the support structure for business counselling,
3. increase knowledge and ability for private and public investments,
4. develop methods for the evaluation of impacts of social innovations,
5. support hubs or network arenas for dissemination of knowledge and research.

As part of the implementation of this strategy, an assignment is given to the Swedish Agency for Economic and Regional Growth to support local and regional initiatives in creating arenas enhancing social innovation and social enterprises [46].

In this article, we will use empirical data from one such project initiative in Southern Sweden to establish an arena for social innovations among local and regional actors. This arena is intended to contribute to greater collaboration, increased employment, more sustainable companies, and solutions to complex societal challenges. Through a number of business loops with presumptive entrepreneurs, the project intends to gather experiences and learning activities to establish a regionally based arena for future development of social innovations.

3. Methodological Framework

The methodological framework intends to problematize how actors are participating in a locally organized project express ambitions, problems, and courses of action for developing social innovations. The intention is to interpret the discourses and arguments that dominate in the implementation of these efforts. To accomplish this, we will use Carol Bacchi's policy theory called What's the Problem Represented to be (WPR-analysis) [23,47,48]. The theory helps to ask critical questions and to challenge axiomatically expressed assumptions in various policies.

Essential to the analysis is to regard societal problems and their solutions as discursively determined with meaning, concepts, and institutionally shaped conditions. In the discourses formed in a policy area, e.g., to develop social innovations, specific forms of conceptual frameworks and institutionally determined understandings shape our practices and working methods in the implementation structure [23]. Thus, the WPR-analysis is a methodological tool to critically ask questions on how public policies are created and implemented. The starting point is that when someone puts forward or suggests something about conditions that are considered to be a societal problem, it is also stated what needs to be solved in a particular activity [47].

The methodological framework combines the WPR-analysis with theories based on analysis of ideas and consists of a sequence in three steps. First, we present the empirical setting and data collection. Second, we introduce the methods based on two ideal types of social sustainability. The ideal types frame the analysis how to map or extract expressions and policy representations in the empirical material. Third, we summarize the approach with questions related to the WPR-analysis.

3.1. Step 1: Empirical Setting and Data Collection

This article will analyze what is represented as a problem when social innovations and social enterprises are created and implemented as a specific policy area in a regional setting. The policy area is manifested through a regional development project co-funded by the European Regional Development Fund and the County Council to support new social enterprises in the southern part of Sweden. The managing authority tasked to select projects into the funding programme and monitoring implementation is the Swedish Agency for Economic and Regional Growth.

The core of the project is to offer coordinated, time-compressed training programs to smaller cohorts of participants who seek to develop social ventures. The program is designed to accelerate the venture development process within a given timeframe by means of enterprise-oriented training, coaching, networking events, and seminars. The typical participants are nascent entrepreneurs, and during the program, they meet 1–2 days per week over ten weeks.

We have used various data sources to account for the many parties involved in characterizing the representation of the policy area. In addition, as the creation and implementation of policy is a process that unfolds over time, we have also collected both retrospectives as well as real-time accounts by relevant parties experiencing the phenomenon of theoretical interest. The data collected enables us to analyze how the involved actors articulate the problems to be solved, what kind of assumptions underpin the representations of the problem, how these representations of the problem come about, and any aspects left unproblematic or silenced.

Our empirical analysis rests on four primary sources of data. First, we have collected documents that provide information about policy intentions. Second, we have followed the planning and implementation of the project from its inception, including continuous changes made in the structure and content of the training program, including being present during different occasions in the training program such as kick-offs, training seminars, guest lectures, and networking events. Third, we have conducted focus group interviews with the project team responsible for carrying out the training programs. Fourth, we have conducted interviews with the social entrepreneurs that have followed the training program. A breakdown of the data used in this study is displayed in Table 1.

Table 1. Data breakdown.

Data Type	Details
Documents	Notes from policy conference, project application, and progress reports.
Participatory observation	Field notes taken to document events, interactions, and artifacts observed in the social setting of the training program.
Interviews with project managers	Two group interviews have been conducted with the project managers during the project period. The project management consist of the CEO of the organization that 'owns' the project, the project manager, a business coach, and the project controller.
Interviews with target group	Interviews has been conducted with 24 social entrepreneurs that followed the training program. The interviews were semi-structured and conducted three months after completion.

3.2. Step 2: Analyis of Policy Representations and Two Ideal Types of Social Sustainability

Analytically, we intend to map expressions in the empirical material that contrasts two ideal types to achieve social sustainability: an individual activation strategy versus a societal equilibrium strategy. The ideal-typical analysis represents a way of doing social science research in a heuristic way, i.e., a process of making abstractions of reality in its purest imaginable form by capturing the essential characteristics of an empirical phenomenon. However, by definition, the ideal type is a reduction of reality that aims to serve as a yardstick or a framework to facilitate comparisons in a constantly changing societal environment [49,50]. When used as a method of comparison, an ideal type enables us to discover the contrasts between ideals and reality. This article will apply two kinds of ideal types that both contain contrasting strategies to contribute to policies for a social sustainable welfare system.

The ideal types in this analysis are based on diverting assumptions or ideas. According to the political scientist Lindberg [51], ideas in public policies or political ideologies consists of three kinds of representations. In this context, Lindberg has developed a conceptualization, which he describes as a VDP triad; (V) as values or value-judgements; (D) as descriptions or judgements of reality; and (P) Prescriptions or practical proposals for action [51]. Lindberg assert that such VDP-triads

> form not only the argumentative, action-guiding and action-directing backbone—the inner structure—of the common ideal-type political ideologies (such as liberalism, conservatism, feminism etc.), but also the manifest or latent inner structure of deliberative political debate, public policy respectively opinion-forming political propaganda (p. 20, [51]).

As with the concept of social innovation, the concept of social sustainability is contested. It is common in the research on social sustainability to problematize the lack of a coherent definition and that the concept has been subordinated to sustainability linked to both ecological and economic development. By utilizing ideal types as a methodological tool we are able to problematize policy representations made by the involved actors in our case.

In the first ideal type—social sustainability as an individual activation strategy—the dominant descriptions concern ideological changes in western societies associated with neo-liberalism and changes in the institutional settings of the welfare state [21], mainly as a change from an emphasis on universal and collective orientations to more individualized and incentive-driven systems. The normative content is based on a bottom-up view where individuals can utilize freedom of choice and a high degree of 'responsibilization'. Public policies should promote models of contractual partnerships between, on the one side, the state and public institutions and, on the other, private actors and organizations in civil society. This means, among other things, policies creating incentive structures, education for lifelong learning, career planning, entrepreneurship, and so on. The policies are directed towards developing strategies for coaching and coping among individuals in the welfare sector [21,52].

In the second ideal type—social sustainability as a societal equilibrium strategy—the leading problematization concerns how the welfare systems are challenged by societal processes such as globalization, digitalization, migration, and urbanization. In general, social development is considered 'wicked' or complex, i.e., that each problem is unique, with no definitive formula, no final solution (rather, processes of trial-and-error), connectivity among several issues, and often based in a local context. The normative principle that guides the strategy is the ability to create an equilibrium between different social values. In one version, [22] used in this article, social sustainability is regarded as an act of balancing three kinds of societal values: justice (distribution of resources as well as inclusion/participation), human development (education, health, quality of life), and security, e.g., crime prevention as well as promotion of social conditions—"proventive security" [22,53,54].

The ideal types displayed in Table 2 will be used as an analytical framework to characterize how involved actors articulate problems in the development process. Firstly, how do the actors express the societal reality that characterizes the project work: Is the interest primarily directed towards overall societal processes, or are there aspects linked to increasing elements of individualization? Second, what normative justifications are expressed by the actors (i.e., on the importance of individual responsibility or norms attached to complex perspectives that weigh in different values)? Thirdly, which policy proposals are advocated by the actors, demands for social policy for coping/coaching among the disadvantaged, or needs for structurally oriented reform processes?

Table 2. Ideal types in efforts to achieve social sustainability.

VDP-Triad *	Social Sustainability as ...	
	... Individual Activation	... Societal Equilibrium
Descriptions or judgments of reality	Descriptions on how changes in the welfare system have occurred from universal and de-commodified social services to more neoliberal elements of individual choice.	Descriptions on how of overall societal processes—globalization, digitalization, migration, urbanization—affects social life.
Value judgments	Social sustainability should be based on individual rights, responsibility, and freedom of choice.	Social sustainability should be based on an integrated approach; an equilibrium among several social values (security, justice, human development).
Prescriptions or practical proposals	Policies for lifelong learning career management, incentive structures, coaching and coping strategies towards individuals.	Policies for structural changes in society concerning participative democracy, liberal arts education, reforms in the health care system, creating "proventive security".

* In this study, we will switch the (V) with the (D) in the sequence and start the analysis with the descriptions, followed by the value judgments, and finally, the prescriptions.

3.3. Step 3: Conclusion—What Is the Problem Represented to Be

Finally, we will conclude the analysis by answering the main question in Bacchi's WPR-analysis. As noted above the task is—in the words of Carol Bacchi—in a 'WPR' analysis is to read

> policies with an eye to discerning how the 'problem' is represented within them and to subject this problem representation to critical scrutiny (p. 21, [55]).

In the critical scrutiny in this study, we will present the main findings and answer the research question on what the problem is represented to be. In the concluding Section 5, we will further problematize the implications of the empirical findings and in the discussion, we will mainly deal with the following issues related to the policy representation in our empirical case:

1. How is the *problem* represented in relation to the two ideal types: on individual activation or societal equilibrium? Is it possible to discern changes in the problem representation among different levels and actors in the implementation process?
2. What kind of character and assumptions prevail (concerning, for example, the policy content, the nature of the policy processes, the design of strategies and working methods)?
3. What are the silences that have been left unproblematic in the representation of the implementation process? Can the problem be thought of differently?

Hence, in the concluding section we will problematize various implications on how to introduce and implement policies supporting social innovations and social enterprises. In the final part, we will present some thoughts on future research.

4. Analysis

In this section, we will analyze how the involved actors articulate issues in the policy process at three levels. Firstly (Section 4.1), we make an analysis at the regional level based on articulations found in policy documents (the national strategy and the regional development strategy), conference documents (material presented at the kick-off conference), and in the application of the development project. Secondly (4.2), we make an analysis at the level of the project team based on articulations made in focus group interviews. Finally (4.3), we analyze the level of the social entrepreneurs based on articulations in the semi-structured interviews. Together, the different levels of analysis provide theoretical and empirical insights into the underlying premises, assumptions, and discourses at play when the policy area is put into practice from creation to implementation.

4.1. The Regional Level—The Initial Formulation of the Policy Problem

The primary impetus behind creating a policy for social innovations and social enterprises in the region can be linked to a publicly funded economic association that provides counselling information, training, and advice for starting cooperative enterprises and supports social entrepreneurship. The organization is structured via independent units in each region all over Sweden. The regional unit of the economic association operating in the southern part of Sweden has, at least since 2010, made several attempts to articulate the need for a comprehensive policy in business development concerning social innovations and social entrepreneurship. Similar to the development for the Nordic countries in general, these ambitions have been met by an interest in the region and some degree of resistance.

In the light of the fact that other counties have taken similar initiatives, the economic association took another initiative in 2016 by inviting potentially interested actors and stakeholders to a kick-off conference. Thus, this initiative was developed mainly by the public actors at the regional level, but a considerable interest was also shown by both for-profit and non-profit participants in the region. A broad range of organizations are invited from all sectors in society. About 50 people signed up, and the conference was attended by representatives from municipalities, state authorities, local action groups (within the EU Leader Program), the Church of Sweden, trade unions, environmental associations, sports clubs, business organizations, and small businesses. The conference included lectures and an exchange of experiences. During the conference, representatives from the economic association documented a rich canvas of needs and challenges for social innovations to address several societal challenges in the region. The presentations at the conference focused on issues such as climate change, work-integrating businesses, rural development, and health care (the documentation, all written in Swedish, can be provided by the corresponding author upon request).

The documentation from the conference forms the basis for continued work to initiate a regional policy. The regional organization led the continued activities in cooperation with the participation from mainly the university and the county council. The policy is concretized in 2017 by initiating a project to build an arena for knowledge, method testing, follow-up research, network building, and business development activities to develop social innovations and social entrepreneurship. The project, which attracted funding from the European Regional Development Fund, started at the beginning of 2018 and lasted until December 2020 (the application, written in Swedish, can be provided by the corresponding author). The intention was to help small businesses combine the logic of entrepreneurship to enhance social values. Examples of activities are:

- creating social innovation labs,
- awareness-raising activities (workshops, teaching conferences, study visits),
- design strategies for commercial development,
- individual coaching of entrepreneurs in starting or consolidating social enterprises.

When analyzing the material produced as a basis for both the conference and the project application, we can conclude that the description of societal development is articulated in terms of dealing with complex social challenges. It concerns issues such as the depopulation of rural areas, growing social exclusion, and climate change. However, some formulations on problems at the individual level, for example, notes on how to create possibilities for individual activation and responsibility. For example, this latter type of problem characterizes a large part of work-integrating social enterprises. The idea is to sell goods and services on the market with companies, often cooperatively organized. The basic idea is that the employees should receive help and support to adapt to working life. These may be people who have had difficulty getting work due to long-term sick leave or due to the integration of refugee immigrants in the labor market. The overall conclusion is that the documentation mainly describes structurally fixated problems. Thus, social innovations are initially motivated as developments in society and concern shortcomings in the societal systems. The development of the new policy area is so far following the strategy at the national level. The main focus is directed towards social sustainability as a societal equilibrium strategy.

Descriptions of the societal development during this initial phase are linked to a number of normative justifications of the policy problem. Several of the speakers at the kick-off conference stressed integrating multi-dimensional perspectives and values to accomplish societal changes. Particularly, this is the case in policies concerning rural development. The capacity to achieve long-term sustainability for rural communities should consider the importance of public services, conditions for private enterprises, infrastructure investments, protection of cultural heritage, and quality of life. This joined perspective also characterizes normative justifications in climate change adaptation that underline processes of integration and coordination both within and across policy areas and organizational levels in society.

Again, to be regarded as a minor part, there are normative values that focus on the responsibility of each individual to reduce various forms of social exclusion. However, it is worth adding that social exclusion is not only analyzed as an individual responsibility but also as a part of shortcomings in the overall social system. Social exclusion is then analyzed as failures in coordinating social investments, for example, housing refurbishment, infrastructure, health initiatives, recreation, and cultural activities. The conclusion is that social exclusion is not primarily due to passivated and benefit-dependent individuals but to poorly developed social institutions.

Against this background, it can be noted that when the new policy is initiated, led by the regional organization, it is explicit that it should primarily address societal changes. However, some elements in the policy initiation emphasize social sustainability based on the responsibilization of individuals, but these are subordinate. The dominant part concerns the lack of institutional arrangements, resources, and support structures to support individuals and organizations that want to invest in social innovations. The policy problem represents intentions to create an institutionalized arena to change and improve conditions for positive social change. One essential part of the policy idea is to give public attention to a bias in that business counseling systems often disregard social innovation and social enterprises.

Consequently, a basic principle in the policy idea is to assert the particularities that characterize positive social change in business policy. The particularity consists of the fact that social innovations aim to achieve structural conditions concerning long-term sustainability. Again, we can note that these conclusions, articulated in the conference materials and the project application, are linked to the strategy at the national level—both in creating an arena and strengthening efforts for active counselling supporting initiatives in starting social enterprises.

If we summarize the analysis so far, we can conclude that the original policy representation is based on an idea to strengthen the institutional capacity in the region to support sustainable forms of social innovations and social entrepreneurship. Although there are activities in the development project that aim to assert a commercial focus and also the need for individual activation, the policy problem is mainly represented to be an issue anchored in a societal context.

4.2. The Level of the Project Team

When the development project started in January 2018, a project team was formed: a Project manager, a Business Coach, and an Administrator. It is also worth noting that the Project Owner (as a director for the economic association responsible for the project) takes an active part in planning and implementing the project as a regular discussion partner to the project team. During the focus group interviews with the project team (including the project owner), a partially changed representation of the problem could be identified. The main focus is gradually displaced from a societal orientation to growing attention on problems associated with the individual entrepreneur and the participants' entrepreneurial mindset.

Thus, we can see tendencies towards a policy shift that become clear in two respects and can also be described as policy dilemmas. Firstly is a dilemma linked to how the project team selects participating entrepreneurs. When the project team describes the selection process, the ambition is to find a segment of entrepreneurs who do not usually seek out the existing support structure. This mainly applies to local business offices, regional actors, incubators, and institutions for the supply of venture capital.

This means that the project primarily addresses a segment of entrepreneurs who have a weak interest and scarce resources in creating new businesses. It should be said that this is a well-considered and strategic choice in the project. Characteristically, this implies that many of the selected entrepreneurs have difficulties accessing various forms of institutional support. Although the project team is aware of the importance of maintaining a clear societal orientation in the project, efforts are required to train the recruited participants for individual responsibility and changed attitudes. The project team has to spend a lot of time and resources on coaching individual participants to develop their social enterprising efforts with a commercial and market-oriented mindset. The risk is that this will lead to a policy shift; the policy problem is to a lesser extent directed towards sustainability in a societal perspective and increasingly towards problems linked to each entrepreneur.

Secondly, we can identify a policy shift in a dilemma associated with the tools and methods utilized in project-making business counseling activities. Often, these tools have been developed to suit traditional business counselling and are not specially designed for developing social businesses. For example, one of these tools or models—called the Business Development Matrix—is sequentially based on business development, emphasizing goal-oriented considerations of the individual entrepreneur and his/her relations to the market, the customers, budgeting, and business acumen.

The project team has made conscious attempts to adapt the tools to managing social enterprise and launched several modifications for this purpose. However, existing tools for business development are foremost based on traditional entrepreneurship. They are subsequently modified to contain at least some aspects to consider social or societal elements. However, it is not the other way around, i.e., that social enterprise forms the basis and is then modified with elements of commercially and individually oriented activities. The difficulties of adapting the available tools for business development with a social and system-changing purpose have been a central challenge for the project team. In addition, in this part, there is a risk that the tools contribute to a shift of the policy representation from a societal focus to thinking characterized by the activation and responsibility of the individual entrepreneur.

The members of the project team express an intention to work for an institutionalized arena for the development of social innovations and social enterprise in the region. In this part, the argument is mainly taken from the project application. The arguments emphasize a need to create structure and visibility for this work in the future. Thus, the project team has worked intensively to establish this arena, mainly to find a venue or place as a unifying base for the activities. The project team claims the need for a common entrance for actors with ideas about developing social innovations. As the work with the project has started, it has become even more evident that this kind of counselling is dependent on personal meetings in real life. Several of the members of the project team emphasize the importance of the exchange of experience on-site in an everyday context. The need to create an institutionalized arena is also motivated because it can be a collective resource for disseminating information, conducting learning activities, and documenting experiences. The argument for the arena is thus linked to the uniqueness that characterizes social innovations compared to other kinds of business counselling.

4.3. The Level of the Social Entrepreneurs

Finally, the views of the participating entrepreneurs have been analyzed. A total of 35 participants have received support through the initiatives implemented in the development project. Therefore, we can categorize the participating entrepreneurs into two main categories. The first, which dominates (about 80% of all), focuses exclusively on individuals as a target group. The second category may also have individuals in focus but have a more substantial element of a societal orientation in their entrepreneurship ideas compared to the first category.

In the first category, where many of the business ideas are similar to each other, the description of reality is fixed on the health and well-being of individuals. The policy problem to be solved is developing attitudes and tools to change people's life situations through individual activation. The tools are intended to influence responsibility and interest in changing their own lives, becoming more harmonious as human beings, dealing with drug problems or gambling addictions, getting out of destructive relationships, etc. The characteristic of this group of entrepreneurs is that they have difficulties describing the significance of their business idea from a societal or social perspective. During the interviews, one continuously returns to challenges in the individual context; therapeutic methods, coping linked to stress management, mindfulness, individual rehabilitation, cultural experiences, etc. The interviewed entrepreneurs have in-depth knowledge and competence in their specialization, often highlighting their personal experiences of the specific problems they want to work on within their enterprise. However, the participating entrepreneurs often lack insights and skills in running a business and conducting it in a business-like way. Commitment is strong on the issues, but the ability to write a business plan, create long-term financing, and manage marketing is not as well developed. The observations made by the project team are that the entrepreneurs have difficulties in running companies with financial viability also appears in our interviews.

In the second category of participants, where we can discern a societal or social orientation, the description of reality is based on a given societal problem. These statements are mainly linked to climate issues in our material, but individual entrepreneurs point to shortcomings in the food supply, issues concerning developing countries, and depopulation in rural areas. In this category, the need for social innovations is justified from a structural perspective and ideas about long-term societal development. However, it should be noted that these constitute a minority among participating entrepreneurs and that there is also a connection among them to individual activation and responsibilization. In this group, however, we can find a more developed and well-thought-out perspective on what is meant by social change compared to how it is in the first category.

We can conclude that there is a minority who intends to develop their future business with a dominant normative notion of being able to influence societal structures among the participants. The dominant group of entrepreneurs is driven by perceptions that primarily contribute to the development of the individual. The participants generally find it difficult to see themselves as "social" entrepreneurs. Instead, they regard themselves as committed human beings with ambitions to run a business. When a large proportion of the participating entrepreneurs work to help individuals get out of their problems, a dilemma immediately arises about running a profitable company at all.

This kind of problem is linked to the type of entrepreneurship that should be the target group to stimulate social innovation and social enterprise. When selecting potential participants who has a predominance of entrepreneurship linked to individual activation, there is a risk of losing the societal dimension. This can then be related to the selection problem that was touched on above about the strategy of the project team to find entrepreneurs who do not fit well into the established support and advisory system for business development in the region.

Regarding the question of what the entrepreneurs want in terms of policy proposals in the future, attention is drawn to the need to continue to receive support to develop their companies. The interviewed entrepreneurs make several suggestions. Many articulate the importance of building networks and, together with other entrepreneurs, create a common platform for cooperation on a reciprocal basis. The project experiences are generally valued positively as they have entered into a collaboration and exchange of experience with other entrepreneurs. Even if you work with different subject areas and varying forms of enterprises, many problems are common. Many interviewees would like to see an established arena for continued exchange and opportunities for continuous advice. There is a call for support to apply for project funding, develop their marketing, get help with contact-creating with other support organizations, etc. The interviewees highlight various needs for support measures. Undoubtedly, the group of active entrepreneurs is interested in developing an arena for social innovation in the region. The interviewed participants emphasize that the arena has a task primarily to support the entrepreneurs in practical parts, such as financing, marketing, and exchange of experience.

4.4. Concluding Analysis

In this section we have answered the three sub-questions that was raised in the introduction, which is summarized in Table 3:

- In what ways are societal reality characterized in the policy representation?
- What normative assumptions underpin the representation?
- How has the representation of the problem been transformed to policy proposals?

Table 3. Main findings from analysis.

Level of Analysis/Material	Descriptions of Reality	Normative Justification	Policy Guidelines
The regional level - pre-evaluation report - kick-off conference - project application	Overall societal processes such as urbanization, climate change, integration, unemployment.	A society that is able to utilize social commitments in civil society in developing public policies for social change.	The need for an institutional arena to create identity and to visualize the specificity of social innovations. A need to weigh in between social sustainability and social investments.
The project team - focus-group interviews	Identification of a segment of individuals and organizations in society that get no attention in public policies for business development.	To expand the opportunities for social change through ideas about entrepreneurship and social innovations	The need for an institutional arena to create networks to enhance business counselling especially among social entrepreneurs
Social entrepreneurs - semi-structured interviews	Focused on problems of social exclusion among individuals.	Social change is dependent on the degree of conscious and responsible activities among individuals. Individual activation *per se*.	The need for an institutional arena to support newly started enterprises in practical issues concerning economic strategies

For all three questions, we can note critical differences between the three levels of analysis. First, on descriptions of reality, different views are articulated between all three levels. Articulations made at the regional level on overall societal processes are not exposed among the articulations made by the project team or among the participating entrepreneurs. The project team is mainly occupied by descriptions of which actors usually do not utilize public business counseling policies. The project team identifies several aspects of society that substantially restrain the options for potential entrepreneurs with social ambitions to start businesses. The policy shift is due to the task given to the project team, namely, to support the ability of individual entrepreneurs to create new social businesses.

Among the participating entrepreneurs, the perspective is, to a large extent, linked to the difficulties individuals may encounter in society. As a result, the descriptions are almost exclusively fixed to social problems among individuals. With some exceptions, the interviewees are not expressing any views that problematize overall societal processes, such as the depopulation of rural areas or structural unemployment, to explain existing social problems among individuals.

In the second category, we can find similarities between the regional and project teams on normative justifications. The articulations circle around the need to utilize and visualize ideas and commitments to develop social innovations in society. Social involvement in society should be linked to policy ambitions in creating entrepreneurship and business development. On both levels, the actors are well informed on social innovation as a globally disseminated policy concept. In important respects, the articulations are inspired by national as well as global initiatives. However, there is an entirely differing normative justification among the participant entrepreneurs, entirely based on individual activation. Undoubtedly, this is the most apparent policy shift in the analysis. The normative understanding of developing social innovations is not rooted among the participating entrepreneurs.

In the third column on policy guidelines, there is a difference in articulations between, on the one hand, the regional policy level and, on the other, the project team and the participating entrepreneurs. At the regional level, the policy guidelines focus on establishing an arena that could be recognized and visualized in the regional policy context. The policy idea consists of asserting the importance of social innovations as a part of strategies in the regional development policy in the region. It should be seen as a political proposal in the region to introduce social sustainability as an area of innovation. At the project team level and among participants, the policy guidelines proposed assert the need for practical tools to create networks among social businesses, seminars, learning activities, etc.

Finally, the analysis findings indicate significant changes in the policy representation when moving from the top to the bottom. The conclusion is that the intentions formulated at the regional level and in the project application have become something partly different during the implementation of processes and activities at the lower levels in the implementation chain. In the concluding section below, we will delve a little further into this conclusion.

5. Discussion and Conclusions

In policy declarations from governments and influential NGOs worldwide, social innovations are seen as necessary and useful ways of managing most of the contemporary social challenges in our common world. This article has turned into the microcosm of creating social innovations and supporting social enterprise in a small region in Sweden. Even in this microcosm, great hopes are attached to social innovations that will address social challenges in society.

During the work on this study, we have continuously been met by a strong commitment among participating entrepreneurs to personally contribute to creating improved living conditions for socially disadvantaged people. As several researchers have pointed out, the policy area is characterized by a distinctive and sometimes conflict-ridden confusion of business acumen as well as well-founded humanism [18]. Thus, we agree with the conclusion made by several scholars that we need to make in-depth analyzes of what characterizes the individual entrepreneurs in their efforts to develop social enterprises [26].

Our analyses provide an in-depth understanding of several critical issues in international research on social innovations. In our study, activities in the policy area are part of a hybridized organizational field between public, civil, and private sectors. Often, the development processes contain ideologically based value conflicts. The main conclusion is that we have found that basic policy values are shifted in the policy process's different phases. We have also shed light on shifting attitudes concerning organizing and developing the support structure for social innovation at the regional level. Thus, issues related to policy processes rather than policy content tend to dominate during the implementation. In the following, the intention is to summarize the analysis and answer the research questions.

Then, what is the problem represented to be in our case? Our analysis indicates that the original problem representation shifted as the activities in the project were implemented. During the initial phase, the policy idea consisted of well-considered ambitions to create long-term sustainable development. However, during the implementation of the project, the problem's representation changes gradually in the direction towards individual activation and responsibility. This process is driven by pragmatic difficulties of defining the policy area, problems of separating means from ends, and the need to make decisions based on a limited range of information and analysis to get things going. Hence, at the project level, the gradual change in coping with the contradictions and multiple complexities facing activities in a dynamic and complex policy area. In this respect, the reformulation of the problem enables the project team to meet project goals and produce legitimacy towards beneficiaries. However, it also means that ambitions expressed in original policy representations become altered where calls for more profound change in institutional regimes transform into more adaptive, performance-related social innovation efforts [20].

The reformulation of the problem occurs when strategies concerning the selection of entrepreneurs for the activities in the project are formulated. Current policies supporting innovation in the region are identified as having a weak support structure for entrepreneurs who have a solid social commitment but lack the abilities and knowledge to start and run businesses. Therefore, the developed strategy is to prioritize entrepreneurs who are not usually part of the support structure within business and growth policies. Thus, the entrepreneurs who participate in the project are driven by business ideas that, in many cases, lack immediate commercial potential and are often strongly linked to personal interest and commitment.

In sum, the conclusion in this part indicates that the policy area has so far been weakly institutionalized. The involved actors in the implementation structures are given high discretion to design their principles and working methods. As a result, policy intentions and decisions weakly guide those who implement the policy.

5.1. Implications

Our study has implications in a number of areas. The findings illustrate the potential friction created when deeply embedded normative principles connected to collective orientations and social welfare meet the individualized and incentive-driven systems that characterize the support structures for fledging entrepreneurs. In this process, the project receives a focus that emphasizes both individual coaching among the involved entrepreneurs and a commitment among the entrepreneurs to support rehabilitation among individuals. This implies that the original policy ambition to adopt market mechanisms to long-term social sustainability principles is silenced in favour of social sustainability as individual activation in a market perspective. In other words, the implementation process tends to repeatedly emphasize principles that social sustainability should be developed using market mechanisms rather than the other way around—that the market economy needs an underlying logic based on long-term social sustainability. In this respect, our analysis pinpoints the "mission impossible" at the project level and where the pragmatic challenges of the social ventures favour an individual activation strategy.

Another kind of silenced policy representation concerns the idea of creating a regional arena promoting social innovations and social enterprises in the region. The original policy idea was formulated as an arena to institutionalize and visualize a long-term political commitment to developing a new policy area. Following the policy shift noted above, the arena is gradually redefined to become a specific site to deal with practical problems related to the needs of the individual entrepreneur—a kind of hub and lab for managing everyday problems. It also means that the arena is losing its political significance as a negotiation network among leading actors from the public, private, and civil sectors in the region.

For example, our participatory observations in this study give a strong impression that there is a lack of consideration among leading actors on what is social in the policy representation. There are few or no conflicts over the formulation of the representation. Moreover, we have so far been able to notice any debate in public spaces or the media. The political parties are absent in the development processes taking place around our case. Correspondingly, significant influence has been placed on officials and experts in implementing the policy.

Moreover, the implementation of projects and measures needs to assert the social dimension in the policy area. Social enterprising efforts are closely linked to processes of restructuring welfare states who are under the pressure to innovate to meet increasing demands of social services [14,56]. It is, however, clear from the empirical material that there are several different perceptions. The selection process, methods/tools, and business plans deal with making visible the issues of wicked problems, institutional structures, and balancing various aspects holistically. Few participants put heavy emphasis in addressing entrepreneurial activities from a societal perspective or supporting ideas of a social sustainable society. The analysis presented in this article thus stresses the need to clarify what is meant by the prefix 'social' in the policy area.

The depoliticization of the implementation process could create a policy representation that is foremost a rhetorical or symbolic figure whose primary purpose is to build a consensus among leading policy actors. The conditions for establishing an institutionalized implementation structure are weakened, and the initiated activities, which are often valued positively by project owners and participants, risk running out and coming to nothing after the project ends. If one refers to anything other than individual activation when developing social sustainability policies, this should include a more activated political leadership. In this case, creating an institutionalized arena is a window of opportunity for building long-term continuity in the implementation processes and as a political forum for policy development with a longer time perspective.

5.2. Limitations and Future Research

We need to acknowledge the methodological limitations in this study. The method used in this study, i.e., qualitative analysis in one case, belongs to the category that has dominated the research field. Because of the dynamic and complex nature of the area, there is a need for broadening the methodological "toolbox" by theory development and the use of multiple, mixed, and iterative empirical methods in the future. [57–59]. Moreover, our study relied on interviews and participatory observations as the primary method of data collection. Even if this has enabled us to stay close to the lived experience of the informants, we also acknowledge the risk of our data being susceptible to social desirability bias. Adding to this, the study is conducted in Sweden, a country that combines a solid tax-funded welfare system with a relatively well-developed support structure for aspiring entrepreneurs. Thus, comparative analyses on policy implementation in different socio-economic and political contexts using cases from other countries are recommended to corroborate and contextualize our findings.

Finally, we make a brief note on issues requiring further research and investigation. Along this line, there is a need for further research to deepen the understanding of how strategies for social innovations and social enterprise include issues concerning how the social dimension is defined. Not least, there is a lack of studies of how the policy area can be linked to ambitions for social sustainability [60–62]. Moreover, as has been mentioned above, we see great potential in comparative research designs by conducting studies of multiple social innovation projects across different regional settings. This would be particularly useful for understanding and explaining how the broader policy context influences the implementation of social innovation projects. Finally, we encourage studies with even longer time spans. Creating and implementing weakly institutionalized policy areas can be a slow and incremental process. However, such initiatives may over time contribute to policy learning that gradually increase the systemic understanding of the implementation and effects of project based and other coordinated efforts to meet social challenges in the region. Following how this learning process emerges and unfolds at various levels would provide great contributions to our knowledge of social innovations as a policy area.

Author Contributions: Conceptualization, J.J. and J.G.; methodology, J.G. and J.J.; writing—original draft preparation, J.J. and J.G.; writing—review and editing, J.G. and J.J. Both authors have read and agreed to the published version of the manuscript.

Funding: This research was funded by The European Regional Development Fund.

Informed Consent Statement: Informed consent was obtained from all subjects involved in the study.

Data Availability Statement: The data presented in this study are available on request from the corresponding author.

Conflicts of Interest: The authors declare no conflict of interest.

References

1. OECD. *Fostering Innovation to Address Social Challenges*; OECD: Paris, France, 2011.
2. European Commission (EC). *Social Innovation Research in the European Union. Approaches, Findings and Future Directions—Policy Review Brussels: D-G Research and Innovation*; EC: Brussels, Belgium, 2013.
3. European Commission (EC). *Social Enterprises and their Ecosystems: Developments in Europe*; EC: Brussels, Belgium, 2016.
4. Nicholls, A. Social Enterprise and Social Entrepreneurs. In *The Oxford Handbook of Civil Society*; Oxford University Press: Oxford, UK, 2011.
5. Barlagne, C.; Melnykovych, M.; Miller, D.; Hewitt, R.; Secco, L.; Pisani, E.; Nijnik, M. What Are the Impacts of Social Innovation? A Synthetic Review and Case Study of Community Forestry in the Scottish Highlands. *Sustainability* **2021**, *13*, 4359. [CrossRef]
6. Mulgan, G. *Social Innovation: How Societies Find the Power to Change*; Policy Press: Bristol, UK, 2019.
7. Grimm, R.; Fox, C.; Baines, S.; Albertson, K. Social innovation, an answer to contemporary societal challenges? Locating the concept in theory and practice. *Innov. Eur. J. Soc. Sci. Res.* **2013**, *26*, 436–455. [CrossRef]
8. Howaldt, J.; Kaletka, C.; Schröder, A.; Zirngiebl, M. (Eds.) *Atlas of Social Innovation: 2nd Volume: A World of New Practices*; Oekom Verlag: München, Germany, 2019.
9. Kerlin, J.A. Defining Social Enterprise Across Different Contexts: A Conceptual Framework Based on Institutional Factors. *Nonprofit Volunt. Sect. Q.* **2013**, *42*, 84–108. [CrossRef]
10. Komatsu, T.; Deserti, A.; Rizzo, F.; Celi, M.; Alijani, S. Social Innovation Business Models: Coping with Antagonistic Objectives and Assets. In *Finance and Economy for Society: Integrating Sustainability*; Emerald Group Publishing Limited: Bingley, UK, 2016; Volume 11, pp. 315–347.
11. De Pieri, B.; Teasdale, S. Radical futures? Exploring the policy relevance of social innovation. *Soc. Enterp. J.* **2021**, *17*, 94–110. [CrossRef]
12. Van der Have, R.; Rubalcaba, L. Social innovation research: An emerging area of innovation studies? *Res. Policy* **2016**, *45*, 1923–1935. [CrossRef]
13. Edwards-Schachter, M.; Wallace, M.L. 'Shaken, but not stirred': Sixty years of defining social innovation. *Technol. Forecast. Soc. Chang.* **2017**, *119*, 64–79. [CrossRef]
14. Defourny, J.; Lars, H.; Pestoff, V. *Social Enterprise and the Third Sector: Changing European Landscapes in a Comparative Perspective*; Routledge: New York, NY, USA, 2014.
15. Gawell, M. Activist Entrepreneurship: Attac'ing Norms and Articulating Disclosive Stories. Ph.D. Thesis, School of Business, Stockholm University, Stockholm, Sweden, 2006.
16. Jensen, H.; Björk, F.; Lundborg, D.; Olofsson, L.E. *An Ecosystem for Social Innovation in Sweden: A Strategic Research and Innovation Agenda*; Institute for Educational Sciences, Lund University: Lund, Sweden, 2014.
17. Boudes, M. Social Innovation: From hybridity to purity through ambivalent relations to institutional logics. *Acad. Manag. Proc.* **2016**, *2016*, 15564. [CrossRef]
18. Mulgan, G.; Tucker, S.; Ali, R.; Sanders, B. *Social Innovation: What It Is, Why It Matters and How It Can Be Accelerated*; Saïd Business School: Oxford, UK, 2007.
19. Lundgaard Andersen, L.; Gawell, M.; Spear, R. (Eds.) Social Entreprenuership and Social Enterprises in the Nordics: Narratives Emerging from Social Movements and Welfare Dynamics. In *Social Entrepreneurship and Social Enterprises: Nordic Perspectives*; Routledge: New York, NY, USA; London, UK, 2016.
20. Perlik, M. Impacts of Social Innovation on Spatiality in Mountain–Lowland Relationships—Trajectories of Two Swiss Regional Initiatives in the Context of New Policy Regimes. *Sustainability* **2021**, *13*, 3823. [CrossRef]
21. Dahlstedt, M. Discourses of employment and inclusion in Sweden. In *Social Transformations in Scandinavian Cities: Nordic Perspectives on Urban Marginalization and Social sustainability*; Johansson, E.R., Salonen, T., Eds.; Nordic Academic Press: Lund, Sweden, 2015.
22. Abrahamsson, H. The great transformation of our time. Towards just and socially sustainable Scandinavian cities. In *Social Transformations in Scandinavian Cities: Nordic Perspectives on Urban Marginalization and Social Sustainability*; Johansson, E.R., Salonen, T., Eds.; Nordic Academic Press: Lund, Sweden, 2015.
23. Bacchi, C. Policy as Discourse: What does it mean? Where does it get us? *Discourse: Stud. Cult. Politics Educ.* **2000**, *21*, 45–57. [CrossRef]

24. Choi, N.; Majumdar, S. Social entrepreneurship as an essentially contested concept: Opening a new avenue for systematic future research. *J. Bus. Ventur.* **2014**, *29*, 363–376. [CrossRef]
25. Ferreira, J.J.; Fernandes, C.I.; Peres-Ortiz, M.; Alves, H. Conceptualizing social entrepreneurship: Perspectives from the literature. *Int. Rev. Public Nonprofit Mark.* **2017**, *14*, 73–93. [CrossRef]
26. Saebi, T.; Foss, N.J.; Linder, S. Social Entrepreneurship Research: Past Achievements and Future Promises. *J. Manag.* **2019**, *45*, 70–95. [CrossRef]
27. García-Jurado, A.; Pérez-Barea, J.J.; Nova, R. A new approach to social entrepreneurship: A systematic review and meta-analysis. *Sustainability* **2021**, *13*, 2754. [CrossRef]
28. Phills, J.A.; Deiglmeier, K.; Miller, D.T. Rediscovering Social Innovation. *Stanf. Soc. Innov. Rev.* **2008**, *6*, 34–43.
29. Littlewood, D.; Khan, Z. Insights from a systematic review of literature on social enterprise and networks. *Soc. Enterp. J.* **2018**, *14*, 390–409. [CrossRef]
30. Littlewood, D.; Holt, D. Social Entrepreneurship in South Africa: Exploring the Influence of Environment. *Bus. Soc.* **2018**, *57*, 525–561. [CrossRef]
31. Murray, R.; Caulier-Grice, J.; Mulgan, J. *The Open Book of Social Innovation*; Social Innovators Series; The Young Foundation: London, UK, 2010.
32. Austin, J.; Stevenson, H.; Wei–Skillern, J. Social and Commercial Entrepreneurship: Same, Different, or Both? *Entrep. Theory Pract.* **2006**, *30*, 1–22. [CrossRef]
33. Mair, J.; Martí, I. Social entrepreneurship research: A source of explanation, prediction, and delight. *J. World Bus.* **2006**, *41*, 36–44. [CrossRef]
34. Hull, C.E.; Lio, B.H. Innovation in non-profit and for-profit organizations: Visionary, strategic, and financial considerations. *J. Chang. Manag.* **2006**, *6*, 53–65. [CrossRef]
35. Ezponda, J.E.; Malillos, L.M. A Change of Paradigm in the Study of Innovation: The Social Turn in the European Policies of Innovation. *Arbor Cienc. Pensam. Cult.* **2011**, *187*, 1031–1043. [CrossRef]
36. Massey, A. Governance: Public governance to social innovation? *Policy Politics* **2016**, *44*, 663–675. [CrossRef]
37. Parrilla-González, J.; Ortega-Alonso, D. Social Innovation in Olive Oil Cooperatives: A Case Study in Southern Spain. *Sustainability* **2021**, *13*, 3934. [CrossRef]
38. Lundgaard Andersen, L.; Hulgård, L. Social entrepreneurship: Demolition of the welfare state or an arena for solidarity. In *Social Entrepreneurship and Social Enterprises: Nordic Perspectives*; Routledge: New York, NY, USA; London, UK, 2016.
39. Defourny, J.; Nyssens, M. The EMES Approach of Social Enterprise in a Comparative Perspective. In *EMES European Research Network*; Routledge: London, UK, 2012.
40. Gawell, M. Societal Entrepreneurship and Different Forms of Social Enterprises. In *Social Entrepreneurship: Leveraging Economic, Political, and Cultural Dimensions*; Lundström, A., Zhou, C., von Friedrichs, Y., Sundin, E., Eds.; Springer International Publishing: Cham, Switzerland, 2014.
41. Lundström, A.; Zhou, C.; von Friedrichs, Y.; Sundin, E. (Eds.) *Social Entrepreneurship. Leveraging Economic, Political, and Cultural Dimensions*; Springer International Publishing: Cham, Switzerland, 2014.
42. Dees, J.G. Taking social entrepreneurship seriously. *Society* **2007**, *44*, 24–31. [CrossRef]
43. Kostilainen, H.; Pättiniemi, P. Evolution of the social enterprise concept in Finland. In *Social Entrepreneurship and Social Enterprises: Nordic Perspectives*; Routledge: London, UK, 2016.
44. Gawell, M. *Social Enterprises and Their Ecosystems in Europe—Country Report: Sweden*; Publications Office of the European Union: Luxembourg, 2019.
45. Gawell, M.; Lindberg, M.; Neubeck, T. *Innovationslabb för Social Inkludering*. Ideell Arena: Erfarenheter från Vinnova Finansierade Project. Available online: https://idealistas.se/bocker/kommande-innovationslabb-for-social-inkludering (accessed on 15 July 2021).
46. *Regeringens Strategi för sociala Företag—Ett Hållbart Samhälle Genom Socialt Företagande och Sociala Innovationer*; M.o. Enterprise: Stockholm, Sweden, 2018.
47. Bacchi, C.L. *Analysing Policy: What's the Problem Represented to Be?* N.S.W. Pearson: Frenchs Forest, UK, 2009.
48. Bacchi, C.; Bonham, J. Reclaiming discursive practices as an analytic focus: Political implications. *Foucault Stud.* **2014**, 173–192. [CrossRef]
49. Lundquist, L. *Implementation Steering: An Actor-Structure Approach*; Studentlitteratur: Lund, Sweden, 1987.
50. Barberis, P. Thinking about the state, talking bureaucracy, teaching public administration. *Teach. Public Adm.* **2012**, *30*, 76–91. [CrossRef]
51. Lindberg, M. Qualitative Analysis of Ideas and Ideological Content. In *Analyzing Text and Discourse: Eight Approaches for the Social Sciences*; Boréus, K., Bergström, G., Eds.; Sage: London, UK, 2017.
52. Jacobsson, K.; Hollertz, K.; Garsten, C. Local worlds of activation: The diverse pathways of three Swedish municipalities. *Nord. Soc. Work. Res.* **2017**, *7*, 86–100. [CrossRef]
53. Abrahamsson, H. Cities as nodes for global governance or battlefields for social conflicts? The role of dialogue in social sustainability. In Proceedings of the IFHP 56th World Congress: Inclusive Cities in a Global World, Gothenburg, Sweden, 16–19 September 2012.
54. Marmot, M. Social determinants of health inequalities. *Lancet* **2005**, *365*, 1099–1104. [CrossRef]

55. Bacchi, C. Introducing the 'What's the Problem Represented to be?' approach. In *Engaging with Carol Bacchi: Strategic Interventions and Exchanges*; Bletsas, A., Beasley, C., Eds.; University of Adelaide Press: Aelaide, Australia, 2012.
56. Clark, K.D.; Newbert, S.L.; Quigley, N.R. The motivational drivers underlying for-profit venture creation: Comparing social and commercial entrepreneurs. *Int. Small Bus. J. Res. Entrep.* **2018**, *36*, 220–241. [CrossRef]
57. Kaletka, C.; Schröder, A. A Global Mapping of Social Innovations: Challenges of a Theory Driven Methodology. *Eur. Public Soc. Innov. Rev.* **2017**, *2*, 78–92. [CrossRef]
58. Wittmayer, J.M.; Pel, B.; Bauler, T.; Avelino, F.; De Bruxelles, U. Editorial Synthesis: Methodological Challenges in Social Innovation Research. *Eur. Public Soc. Innov. Rev.* **2017**, *2*, 1–16. [CrossRef]
59. Granados, M.L.; Hlupic, V.; Coakes, E.; Mohamed, S. Social enterprise and social entrepreneurship research and theory. *Soc. Enterp. J.* **2011**, *7*, 198–218. [CrossRef]
60. Rinkinen, S.; Oikarinen, T.; Melkas, H. Social enterprises in regional innovation systems: A review of Finnish regional strategies. *Eur. Plan. Stud.* **2016**, *24*, 723–741. [CrossRef]
61. Asenova, D.; Damianova, Z. The interplay between social innovation an sustainability in the Casi and other FP7 projects. In *Atlas of Social Innovation—New Practices for a Better Future*; Howaldt, J., Kaletka, C., Schröder, A., Zirngiebl, M., Eds.; Sozi-Alforschungsstelle, TU Dortmund University: Dortmund, Germany, 2018.
62. Young, D.R.; Searing, E.A.M.; Brewer, C.V. *The Social Enterprise Zoo: A Guide for Perplexed Scholars, Entrepreneurs, Philan-Thropists, Leaders, Investors and Policymakers*; Edward Elgar Publishing: Cheltenham, UK, 2018.

 *sustainability*

Systematic Review

The Role of Public-Private Partnerships in Housing as a Potential Contributor to Sustainable Cities and Communities: A Systematic Review

Terence Fell [1,*] and Johanna Mattsson [2,3,4]

1 Division of Political Science and Economy, Mälardalen University, 72123 Västerås, Sweden
2 Division of Industrial Economics and Organization, Mälardalen University, 72123 Västerås, Sweden; jomt@du.se
3 Division of Business Administration and Management, Dalarna University, 79188 Falun, Sweden
4 The Future-Proof Cities Graduate School, 80176 Gävle, Sweden
* Correspondence: Terence.fell@mdh.se; Tel.: +46-736620811

Citation: Fell, T.; Mattsson, J. The Role of Public-Private Partnerships in Housing as a Potential Contributor to Sustainable Cities and Communities: A Systematic Review. *Sustainability* **2021**, *13*, 7783. https://doi.org/10.3390/su13147783

Academic Editor: Ingemar Elander

Received: 15 June 2021
Accepted: 10 July 2021
Published: 12 July 2021

Publisher's Note: MDPI stays neutral with regard to jurisdictional claims in published maps and institutional affiliations.

Copyright: © 2021 by the authors. Licensee MDPI, Basel, Switzerland. This article is an open access article distributed under the terms and conditions of the Creative Commons Attribution (CC BY) license (https://creativecommons.org/licenses/by/4.0/).

Abstract: Today cities face the increasing negative consequences of the unsustainable course society is set on. Climate change, biodiversity loss and increasing spatial segregation are testament to this. The effects of these issues often exceed the coping capacity of individual urban housing developers. Thus, an antidote to the current neoliberal trend must be found in collaborations such as public-private partnerships (PPP). Here the shortcomings and limitations of PPP and its potential ability to solve the problem of unsustainable urban development are investigated. Using the Doughnut Economics (DE) model as a general guide, a systematic literature review is conducted. The results reveal evidence that PPPs are unjust and exclude local actors from collaborations. Hence, resident participation and inclusion is considered the best strategy for PPP to evolve as a future guarantor of the sustainable city. First, however, major differences in the character of issues that connect the global model of sustainability to the harsh reality of the local context need to be addressed. This gap concerns the city's social foundation and ecological ceiling. The DE model applied herein is an excellent tool to test the scope and depth of local collaborations such as PPPs and reflect on international treaties such as SDGs.

Keywords: public-private partnership; Nordic; governance; housing; future proof cities; sustainability; urban development; Doughnut Economics; sustainable city

1. Introduction

Urban housing developers in today's cities need to better understand the relationship between ecological and social sustainability and Public-Private Partnership (PPP), concerning the latter's potential to realize future national policies and international treaties. Cities today are at risk of facing the increasing negative consequences of climate change while they themselves are responsible for 75% of the world's emissions due to excessive energy use [1]. This well-documented problematic is being exacerbated by the inequity of aggressive neoliberal processes such as gentrification and subsequent displacement, fuelling what best can be described as an out-of-control spatial segregation [2–4]. As a double-edged problem, sustainability includes several aspects of urban development in the city. First, on the ecological side of the sustainability coin, challenges, such as energy poverty, bad air quality, noise pollution, waste, excessive consumption, irresponsible land (ab)use, etc., need to be addressed [2–4]. Second, flipping the sustainability coin, the city's social foundation is threatened by a mix of urban processes such as housing, education, health, well-being, social services, governance, cultural heritages, safety and employment [2,3,5]. The point to be made in this review is that these listed challenges to the sustainability of future cities, here defined from indicators and measurements used by scholars like Tanguay et al. [2],

Chan [3], Steffen et al. [4], Raworth [5], often exceed the coping capacity of individual urban developers such as private and public companies, including municipalities [6,7]. Therefore, the city's usual combination of influential stakeholders needs to tackle this complex dual problem of sustainable urban development when they enter collaborations such as PPP, particularly concerning housing development, the focus here [7–9].

Since it is both malleable and "depict[s] the two boundaries—social and ecological—that together encompass a safe and just space for humanity" [5] (p. 48), the Doughnut Economics (DE) model (Figure 1) is one way to either identify the issues that constitute the aforementioned connection (the bridge), or the issues that are missing (the gap). The systematic literature review is not just a way to translate the DE model from the global to the local level of analysis; it also elucidates the strengths and weaknesses of the connection between sustainability and PPP.

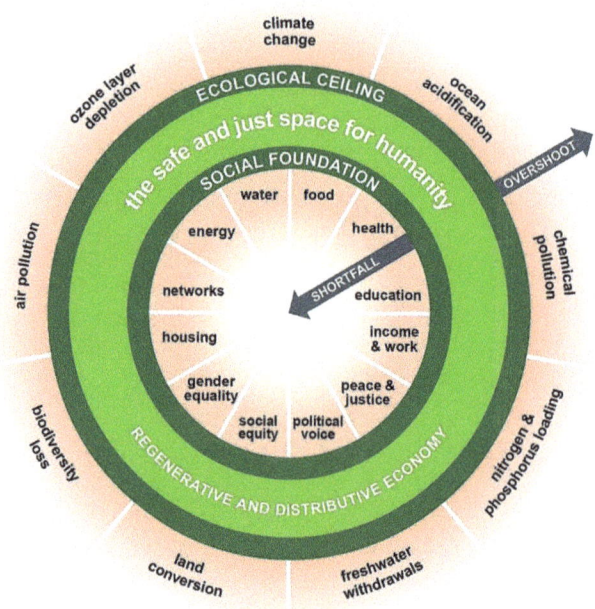

Figure 1. The Doughnut Economics model, reprinted from [5], Lancet Planetary Health, 2017.

The DE model has twelve social indicators derived from internationally agreed minimum standards for human wellbeing such as the Sustainable Development Goals (SDGs) establish in 2015 [5]. The nine ecological indicators refer to the Planetary Boundaries (PBs) developed by Rockström et al. [10] and Steffen et al. [4]. If one of these critical processes that constitutes a PB is overshoot, irreversible changes in the Earth's system are inevitable [4,10]. As mentioned, the main cause of detrimental social development and degradation of PBs originates in cities [11]. This, and the fact that it is used in multiple cities in their transition toward sustainability, for instance, Amsterdam has become a famous example [12], is another reason for choosing the DE model. The DE-model has been adopted by Luukkanen, Vehmas, and Kaivo-Oja [13]; Roy, Basu, and Dong [14]; and Saunders and Luukkanen [15], as a first attempt to develop a method that can be used to compare countries and regions. However, in the application of the DE model herein the authors contribute to this international research by determining the actual scope and characteristics of sustainability efforts in collaborations such as PPPs and by applying it as a broad and new method to guide literature reviews that focus on similar collaborations in local contexts such as municipalities.

Falling partly within the parameters of the UN's sustainable development goal (SDG) 11 to make cities and human settlements inclusive, safe, resilient, and sustainable [2,3], PPP is a recent and growing form of collaboration that to date bridges gaps in infrastructure between essential city services and utilities such as transport, health care, and energy supply [16]. In a literature review of PPP, Hodge and Greve [17] investigated the purpose of PPP and found it to be multifarious and open to interpretation. That is, it can be viewed as everything from a new chapter in privatization to an attempt to measure performance in public sector services. Since the goals of PPP apparently vary, so too does its definition. For this reason, PPP is initially defined broadly as a partnership between the local government (municipality), its administration, and private housing developers [18].

The literature review is also an opportunity to investigate the extent of the gap between studies that focus on PPP, on the one hand [19–25], and studies that focus on sustainability, on the other [4,5,10,11]. Although only a very small proportion (3–4%) of hundreds of articles were written between 2015 and 2021, they can still be used to determine the anatomy of the connection between the PPP in the Nordic housing context and sustainability, the focus of this article. Most international literature today on partnership between public and private actors is either focused on how to improve the partnership per se, often by identifying critical success factors [19–22,25–27] or how it can better manage risk [23,24,28]. Moreover, most of the literature on PPP and sustainability either focuses on countries outside of the Nordic context (see [29–32]), or on sectors other than housing such as waste management, [33], water management [34], and transport [35]. Thus, this systematic literature review is a contribution to this literature and is a first step in deepening the current understanding of the role of PPP as a potential contributor to sustainable development in cities in the Nordic context.

In retrospect, the 25 studies in the review where PPP has been coupled with sustainability in the Nordic housing context reveal criticism toward this kind of collaboration [36–38]. With this in mind, the evidence suggests that, in its present form, PPP enables an asymmetrical power relationship between the municipality and the private sector, on the one hand, and residents, on the other [39]. To illustrate from a Swedish case, in one of Gothenburg's urban frontier neighbourhoods, the Gustaf Dalén neighbourhood, residents and property owners were shown to have no influence over a PPP's plans to redevelop the area and were subsequently forced out [40]. The authors are not discouraged by this example of unsustainable housing development. On the contrary, from this criticism, it is apparent that PPP has the potential to improve the future sustainability of the city's socio-ecological context. This justifies a closer review of the literature to connect the dots between the different issues that traverse the apparent PPP-sustainability gap.

From the scattered body of knowledge concerning PPP, a new understanding of its potential role in facilitating a transition toward a more sustainable housing development in the Nordic context is possible. The authors endeavour therefore to identify, with the support of the DE model, what is missing from the social and the ecological efforts of the current PPP that can be utilized to create a steppingstone in strengthening its future potential. Continuing from previous work on the DE model, the approach applied herein is a normative one, that is, it is designed to identify the shortcomings and limitations of PPP in order to evaluate its potential as a crucial and essential keystone in the sustainable foundation of housing development in the Nordic city context [39,41].

Since collaboration is the norm for current policy implementation, PPP is in a better position than individual urban developers, but not yet sufficient, to bring about a more sustainable housing development [42]. Thus, the need to examine and re-evaluate the PPP is clear. The purpose of this critical literature review is to determine if and how the PPP can achieve a sustainable urban renewal of the future city that appeals to its communities (SDG 11). To this end, it is important to consider how the character of the issues in the DE model change when traversing the global-local divide. While simultaneously exposing their strengths and weaknesses, it is also necessary to identify which of the issues that currently connect the PPP to sustainable development (the bridge), and which issues have the potential to do so (the gap). This is done by applying a novel approach, which is carrying out a systematic literature of PPP in housing development in the local Nordic context to reveal the arguments and themes intrinsic to each concerned issue in the DE model, ultimately augmenting it.

2. Methods
2.1. Systematic Literature Review

A systematic literature review seeks to summarize prior work, extend theories, and evaluate a body of work with a critical lens [43]. Therefore, to build on and advance this theoretical understanding of achieving sustainable housing development, a literature review of a limited body of research that focuses on PPP and current housing development in the Nordic context is conducted. The literature review's protocol is based on a pre-defined structure, intrinsic to the stepwise approach characteristic of the Preferred Reporting Items for Systematic literature reviews and Meta-Analysis (PRISMA) statement [44]. This method was chosen for three reasons. First, it allows an interpretation of the potential role of PPP in housing development in the literature from the perspectives of social and ecological sustainability. Second, it also allows the authors to expose the gaps that can be filled by the PPP. Thus, the PRISMA statements support the research when reporting from the literature on housing development in the Nordic context (see [45] for a similar approach). Third, concerning the issues of reliability and validity, the systematic literature review underpinned by the PRISMA statement ensures reproducibility and replicability of the study [43,46].

2.1.1. Search Strategy in Identification Phase

Sustainability and urban development are two research fields that are interconnected and thus known for being multidisciplinary. In the coming search for relevant knowledge, the authors therefore chose three widely recognized, high quality, and multidisciplinary databases Web of Science Core Collection and Scopus. The search was conducted in March 2021. Keywords and Boolean operators were combined to establish the search for literature. These are "Public Private Collaboration" OR "Public Private Partnership" OR "Governance" AND "urban" OR "housing" OR "cit*" OR "neighbourhood" OR "communit*" AND "Sweden" OR "Nordic" OR "Denmark" OR "Norway" OR "Finland" OR "Iceland". The search was limited to only peer-reviewed scientific journal articles written in English. Since the potential role of PPP is investigated, only articles from the period 2015–2021 were included. In addition, one record was included from previously identified articles. This resulted in 683 identified records, as demonstrated in Figure 2. The main reason for choosing the timeframe 2015–2021 has to do with the fact that the context of the political landscape is rapidly changing. One major change in the political landscape in Nordic countries such as the old welfare state of Sweden is the emergence of neoliberal politics and policies in the late 1990s [47]. The selected timeframe captures the effects of this transformation such as spatial segregation and displacement as they continue to worsen considerably [48]. This ideological transformation, its recent effects combined with an acute need to combat climate change, paints an accurate picture of the double-edged sustainability problematic within which PPPs now operate.

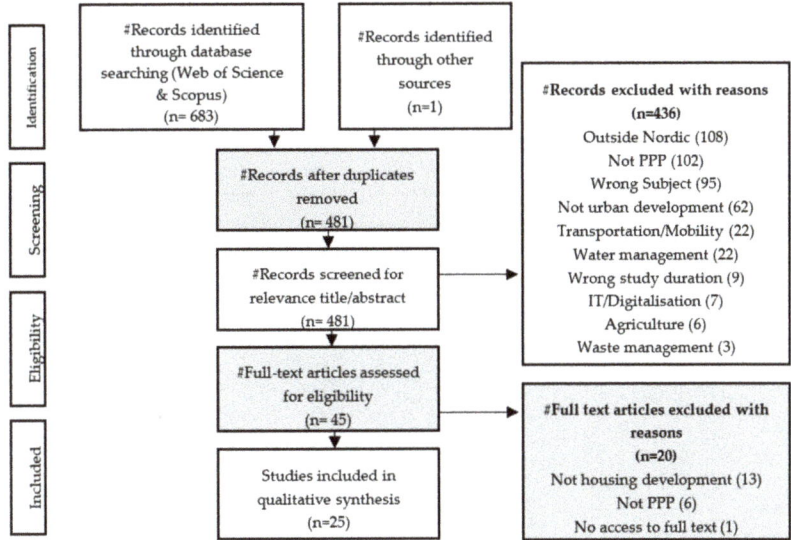

Figure 2. PRISMA statement flow diagram for PPP and urban (housing) development, reprinted from [44], Systematic Reviews, 2015.

2.1.2. Exclusion and Inclusion Criteria in Screening and Eligibility Phase

The retrieved articles were organized in the software Rayyan.ai [49]. In Rayyan, duplicates were removed, and the title/abstracts were screened for relevance. The inclusion criteria in the screening were: Nordic cases (Nordic cases combined with cases outside the Nordic context were excluded) on the topic urban development and PPP. At this juncture, the search was widened by using urban development instead of housing development in order to get a complete picture of the field. In addition, papers on the topic of urban development and/or PPP but concerning a specific discipline not relevant for the review were excluded. For instance, excluded disciplines were water management, waste management, transportation, agriculture, etc. Based on this screening, a refined selection of 45 papers were assessed for eligibility. In the eligibility assessment, one paper was excluded because of difficulty in gaining access to the full text. At this later stage, the authors also narrowed the inclusion criteria to exclude papers that did not combine housing development and PPP. This resulted in a final number of 25 articles to review. It is noteworthy that only 3.6% of all articles in the search pertain to both the PPP and housing development, revealing a narrow connection and probable major gap between the fields of collaboration and urban studies.

2.2. Content Analysis

The content analysis was divided into two phases. Based on the 25 retrieved articles, the first phase started with a screening of the abstracts to find and conceptualize the content of the articles with regards to the issues that constitute the DE model [4,5]. A number of initial themes emerged from common arguments that are used to describe similar phenomena. For example, "access" is a theme found to be intrinsic to the social equity issue and derived herein from the argumentation underpinning phenomena such as "access to urban green space" [50] and "access to affordable housing" [51,52]. In the second phase, by reading the full text of each article in a careful manner [53], the authors reviewed the themes and arguments and then added them with the relevant sources to each issue in table form. In the review of the articles, the authors strived to separate the researcher(s) from the object of analysis. For instance, the themes identified represent what is mentioned in the articles as the focal points in today's housing development and PPP in relation to sustainability. Critical arguments represent the recommendation for the future, its potential, by the researchers in the reviewed articles.

Finally, combining the DE model with articles that critically assess PPP in housing development allows researchers to quickly identify issues that transcend the global-local divide. Those issues that are addressed more frequently are those assumed to be important and, thus, attract most criticism. From this, the shortcomings and potential of local collaborations such as PPP can be identified in relation to any of the UN's SDGs, in this case SDG 11. Furthermore, the arguments put forward by the authors of the articles and sorted into a number of themes reflect on the concerned issues in the DE model. In this manner, it is possible to determine the anatomy of the connection between PPP and sustainability and the extent of the gap that it bridges.

3. Results

The first result is that the authors' reading of the literature that combines the PPP and housing development (see Table 1) found that only seven of the 21 issues touched on by the DE model are covered by the reviewed research. The identified issues are social equity, political voice, justice, social networks, climate change, land conversion, and biodiversity. The 14 issues that are not mentioned in the literature are therefore not included in the results, but some (or all) of these will be reflected on in the discussion. The seven issues are associated with some shortcomings that characterize the PPP and, from the authors' point of view, hamper its ability to achieve sustainability in current Nordic housing development. Four of these connect with social sustainability and two with ecological sustainability, as defined by Raworth [5] (Table 1). Focusing on these issues will bring the PPP closer to achieving SDG 11, expanding beyond what is acceptable within the parameters of economic growth. Just because an issue such as gender equality is not mentioned in the review does not mean that the PPP has already achieved this goal. On the contrary, it is most likely a sign that this issue has not yet reached the drawing board of the PPP. Table 1 reveals the severity of each issue in relation to how many articles, that is, researchers identify it as a problem.

Table 1. The focus of criticism directed at the PPP in current research in relation to the DE model.

Study	Social Sustainability				Ecological Sustainability		Sum
	Equity	Political Voice	Justice	Social Network	Climate Change	Land Conversion and Biodiversity	
Olsson, Brunner, Nordin, and Hansson 2020 [50]	x	x	x	x	x	x	6
Borgström 2019 [54]				x			1
Sørensen & Torfing 2020 [55]		x	x	x	x		3
Bonow and Normark, 2018 [56]		x		x		x	3
Glaas et al. 2019 [57]				x	x	x	3
Hyötyläinen and Haila 2018 [51]	x	x	x			x	4
Lidegaard, Nuccio, and Bille, 2018 [58]	x		x				2
Fors, Nielsen, Konijnendijk, van den Bosch, Jansson 2018 [59]		x		x		x	3
Elander and Gustavsson 2019 [60]			x	x	x		3
Candel, Karrbom Gustavsson, and Eriksson 2021 [61]			x	x			2
Hermelin and Jonsson 2020 [62]			x	x			2
Noring, Struthers, and Grydehøj 2020 [52]	x		x				2

Table 1. Cont.

Study	Social Sustainability				Ecological Sustainability		Sum
	Equity	Political Voice	Justice	Social Network	Climate Change	Land Conversion and Biodiversity	
Puustinen and Viitanen 2015 [63]	x		x	x			3
Valli and Hammami 2021 [64]	x		x				2
la Cour and Andersen 2016 [65]			x				1
Smedby and Quitzau 2016 [66]				x	x		2
Berglund-Snoddgrass, Högström, Fjellfeldt, and Markström 2021 [67]	x		x				2
Juhola, Seppälä, and Klein 2020 [68]		x		x	x		3
Gohari, Baer, Nielsen, Gilcher, and Situmorang 2020 [69]		x		x	x		3
Noring 2019 [70]	x		x		x		3
Thörn and Holgersson 2016 [40]	x		x				2
Schultz Larsen and Nagel Delica 2021 [71]	x		x	x			3
Andersen, Ander, and Skrede 2020 [72]	x		x	x			3
Richner and Olesen 2019 [73]	x		x	x			3
Storbjörk, Hjerpe, and Glaas 2019 [74]			x	x	x		3
Sum	12	7	17	17	9	5	67

What is striking is the asymmetry in the focus of criticism directed at the PPP's housing development. Only 21% of the issues touched on by the literature in review pertain to ecological sustainability (Table 1), but once again not necessarily implying that PPP has achieved these goals. The second result is that PPPs in this study are always being criticized and mostly for their lack of social sustainability, undermining the social foundation of the future city and its communities.

What is also striking, and the third result, is the fact that the two main issues, justice and social networks, touched on by most researchers in the study are those not covered by SDG 11. However, it is important to note that the PPP does not need to be limited by SDG 11 and its subgoals. In fact, PPPs will need to address all the SDGs if they are to tackle the challenges of sustainability in a holistic manner. With the DE model in mind, PPP can in theory transcend the boundary of economic growth by being just and by broadening its social networks (Table 2).

Table 2. Shortcomings of the PPP and its potential effect on SDG 11.

Contentious Issues	Realm	Subgoals SDG 11
1. Social equity (3 themes; 8 arguments) 2. Political voice (2 themes; 6 arguments) 3. Justice (3 themes; 11 arguments) 4. Social networks (2 themes; 8 arguments)	Social	"Access to adequate, safe and affordable housing and basic services as well as inclusive green and public spaces" "Capacity for participatory, integrated and sustainable human settlement"
5. Climate change (3 themes; 9 arguments) 6. Biodiversity and Land Conversion (3 themes; 5 arguments)	Ecological	"Policies and plans towards inclusion, resource efficiency, mitigation, adaptation and resilience to disasters" "Efforts to protect and safeguard the world's natural heritage"

As it stands, the subgoals of SDG 11 can be sorted into the realms of both sustainability's social and ecological foundation (Table 2). This implies that the PPP can only contribute to the attainment of the SDG 11 concerning social equity, political voice, climate change, and land conversion and biodiversity. Subsequently, this confines the plethora of identified arguments to the number of issues identified in the review of PPP and urban and housing development (see Sections 3.1 and 3.2 below). However, critics of PPP want it to push beyond the goals of SDG 11, particularly concerning the issues of justice and social networks (as in Table 2). In essence, a new gap in knowledge becomes apparent in a comparison of the DE-model and the subgoals of SDG 11, but only in relation to the city and its communities. The authors are aware that other SDGs deal with justice and social networks in relation to other issues, but not directly in relation to the city and its communities.

It would be easy to view the relationship between SDGs and the DE model as compatible. This is not the case. In fact, the latter transcends GDP growth, and the SDGs do not, leaving room for the PPP to move beyond economic growth. Therefore, the role of the third (economic) pillar of sustainability, which may otherwise seem to be a bit like the elephant in the room, is considered in this review.

To identify its potential as a contributor to the city's sustainability, it is prudent to determine the character, and reveal the content of, the argumentation directed at current PPP in the Nordic context. However, the PPP may not be willing to, or cannot, assert itself to erase what the reviewed research has identified as its transgressions. At this juncture, the reader is reminded that the research goal is a normative one, that is, to determine the PPP's potential as a contributor to sustaining our future cities.

3.1. Socially Sustainable Housing Development

When surveying the social foundation of housing development, it is important to note that although "there are techniques for measuring a reduction or an increase in quantities of CO_2 and for measuring economic gains for a housing company, there are no comparable yardsticks for 'social sustainability', i.e., there is no 'social dioxide' to measure" [75]. Nevertheless, and guided by Raworth [5], four contentious issues have been identified in current research. These outline the future reach of the PPP's potential social sustainability goals, that is, its ability and ambition to engender social equity, be responsive to the collective voice of residents, be a fair developer, and finally, achieve these ends by spinning a wide web of robust social networks. By taking a point of departure in definitions of the four identified social foundation issues used by proponents of the DE model, it is possible at a later stage to compare them with the content and form of the arguments and themes intrinsic to each of these issues as they are systematically described in the literature review.

In Bending Stopper, Kossik, and Gastermanns' [76] version of the DE model in the context of housing development, engendering social equity is first and foremost about housing developers treating different groups of residents equally. Housing companies should, therefore, cooperate socially with local actors in accordance with corporate social responsibility standards. Being responsive to a collective voice is defined as creating the conditions for residents to participate in, or influence, corporate management. Justice is defined in terms of vulnerability and safety. Being a fair developer is, thus, about minimizing residents' vulnerability to housing development. This definition is narrow in comparison with, for example, Jane Jacobs' vision of a just city, which advocates among other things that policy makers are open for anti-subordination [77]. Finally, robust social networks are addressed in terms of generating conditions conducive to a resilient neighbourhood social culture.

Consequently, the fourth tangible result is that the PPP has not yet tackled the full spectrum of the social foundation of sustainability, as depicted by Raworth [5] and Stopper et al. [76] in the DE model. In fact, as it stands, it seems to ignore gender equality, neither does it appear to promote education and guarantee income and work nor cater for the health of residents. Although focus on four issues narrows the scope of the analysis, there is still an opportunity to dig deeper into them by putting each of them under the analytical lens to find the arguments and themes that evoke criticism from colleagues. In essence, putting each issue through a process of softening up, by introducing new contentious themes to gain a new perspective, widens the research community's horizon concerning the potential of the PPP. Identifying the themes intrinsic to each issue is in itself a result, that is, revealing how the PPP could become a solid and essential segment in the social foundation of sustainable cities and communities. Ultimately, the analysis will reveal the sufficiency of the PPP as a necessary contributor to mitigating an unsustainable global and local development.

3.1.1. Social Equity

In all, twelve (48%) of the reviewed research articles pertain in one way or another to the social equity component of the social foundation of sustainability (Table 1), as depicted in the DE model [5]. After determining each of these article's common theme(s), arguments are sorted under three general headings: access, ownership, and implementation (as in Table 3). Each theme is made up of a number of arguments put forward by the author(s) if a PPP is to achieve social equity. All in all, we highlight eight arguments in Table 3 that can consolidate PPP as part of the future city's social foundation.

Table 3. Arguments for the construct of "social equity in housing development".

No.	Themes	Sources	Arguments for Social Equity
1.	Access	[40,50–52,64]	Guarantee the availability of urban green space [50] Increase more affordable social housing via social mixing and positive discrimination [51,52] Avoid landscapes of exclusion and gentrification that widen rent gaps [40,64]
2.	Ownership	[51,58,67,72]	Cultural districts with housing for all citizens [58,67] Avoid building for wealthier homeowners and favouring the preferences of middle and upper classes [51,72]
3.	Implementation	[50,58,63,64,67,70,71,73]	Avoid neoliberal governance of advanced urban marginality [64,71,73] Promote a bottom-up and top-down mixed approach including social services is desirable [58,67] Promote better decision-making processes to void inefficiencies in bureaucracy [63,70]

Under the theme of access, some researchers identify the need for the PPP to make green space more available for resident's irrespective of their class status [50]. Other authors suggest increasing the affordability of social housing [51,52]. Finally, when evaluating entrepreneurial real estate policy in Finland Hyötyläinen and Haila [51] (p. 144) emphasize in the following quote that positive discrimination can increase access to new housing development:

> Helsinki, a small Nordic welfare city, has so far been able to avoid inequalities that generate distress in large European and American cities. This can be explained by referring to a well-functioning social policy, instruments like the production of social housing, and the policies of tenure mix, social mix and positive discrimination.

Ultimately, a policy of social equity that guarantees access to affordable social housing and green space can also contribute to avoiding the now common and ubiquitous processes of segregation that hamper the attainment of social sustainability goals [64]. Another social equity and access hurdle is, for instance, the rent-gaps identified by Thörn and Holgersson [40] in the housing context in Gothenburg, Sweden.

The second aspect of social equity that researchers focus on is ownership. According to Hyötyläinen and Haila [51], a housing development PPP has the potential to avoid building for exclusively only wealthier homeowners. However, this assertion is ambitious since Andersen, Eline Ander, and Skrede [72] (p. 709) show that:

> ... developers are influencing demographic, material, social and cultural changes through their investments and are consciously and strategically reshaping places to increase profits. The profitable 'rent gap' – that is, the gap between the current income earned by a property and possible future income (Smith, 1987) at Tøyen and Grønland – seems to be the driving force for the developers investing in these areas.

For this reason, it is suggested that by building what Lidegaard, Nuccio, and Bille [58] and Berglund-Snoddgrass et al. [67] call cultural districts, the PPP is given an incentive to plan and cater for a wider range of resident and entrepreneur preferences, not just those of the privileged affluent. Lidegaard, Nuccio, and Bille [58] (p. 16) claim that

> governance models should be designed according to policy goals, which are often conflicting, and therefore any proposal for a cultural district should balance equity and efficiency norms to match the expectations of involved stakeholders.

Based on the number of mentions in the literature review, implementation is by far the largest theme in social equity (Table 3) and sheds light on the tendency of the PPP to develop housing and public space within a system of neoliberal governance [71,73]. In the reviewed literature, most researchers argue that neoliberal policies do not resonate well with policies of social equity (as described in Table 3) but are often implemented by inciting fear and anxiety among poor and affluent residents alike. Olsson et al. [50] (p. 311) gives one reason why this can come about:

> This anxiety is not just expressed as fear for increased costs, but also as a long lasting emotional experience caused by having your belongings destroyed and enduring long-lasting renovations.

For this reason, and with the attainment of social equity in mind, the researchers in this study recommend that the PPP apply a mixed, that is, top-down and bottom-up, approach to housing development that includes social services. This kind of implementation is more equitable since it satisfies the preferences of residents from different income brackets as well as a wide range of entrepreneurs, coming to terms with an otherwise inefficient and subsequently socially unsustainable decision-making process [63,67]. In fact, Puustinen and Viitanen [63] (p. 495) indicate

> that the decision-making process is unestablished, and challenges exist on three levels: (1) legal and land use planning, (2) collective action and management and (3) required professionals. These issues need to be considered in order to develop better practices for the process, and also, when assessing the feasibility of infill development for housing companies from the land use planning, legal and economic perspectives.

In sum, researchers suggest that new planning perspectives that include residents' and developers' preferences ought to be adopted by the stakeholders that constitute the PPP if future housing development is to be built upon a solid foundation of social equity. They imply that this will not be possible if the PPP continues to rely on current neoliberal justifications.

3.1.2. Political Voice

Seven (28%) of the reviewed research articles touch on the issue of residents' collective (community's) political voice (Table 1). In other words, political voice is also a piece, albeit a smaller one, of the social foundation pie than, for instance, social equity is. To reiterate, from the reading political voice can be sorted into two predictable themes: participation and citizenship (as in Table 4). While participation is a civic culture phenomenon, that is engaging residents and communities in local issues, citizenship is more focused on the rights of residents, that is, the need to be heard, included, and organized. Together, the authors of these articles argue that if the PPP listens to the political voice of residents and local communities, it will benefit their housing development and make it more socially sustainable.

Table 4. Arguments for the construct of "political voice in housing development".

No.	Themes	Sources	Arguments for Political Voice
1.	Participation	[50,59,68,69]	Promote participatory structures [50] Promote participatory culture [50,59,69] Promote participating in collaborative initiatives [50,59,68]
2.	Citizenship	[50,51,55,56]	Give residents a louder voice [50] Guarantee inclusion of all concerned citizens [51,55] Promote citizen led initiatives [56]

Looking through the analytical lens used here, it is obvious that political voice often pertains to the establishment of structures that engender a culture of resident participation [50,59,69], as well as constitutes the basis for possible joint collaborative initiatives between residents and the PPP [50,59,68]. An example from Sweden shows how diverse and inclusive a participatory structure can be in terms of stakeholder involvement Olsson et al. [50] (p. 310):

> Some of these structures concern interactions between different property owners, for example the BID [Business Improvement District] and local divisions of the Swedish Union of Tenant Association, as well as between property owners and their tenants.

In this case, the concerned authors are highlighting the possibility of building on existing networks that already include resident participation, not just PPP stakeholders. However, according to Juhola, Seppälä, and Klein [68], there is still room for much improvement. They [68] (p. 24) say that there should be more

> ... emphasis on creating innovative solutions in partnership with the private sector and a focus on efficiency has disturbed the long-term horizon of urban planning and democratic legitimacy, which are both resource and time demanding.

There can, therefore, be resistance within the PPP to new ways of thinking. The PPP needs incentives such as a more democratically legitimate role in future urban planning. This can redirect its focus towards laying a more solid social foundation that, in turn, contributes to sustaining the city and its communities.

Concerning citizenship, and depicted in Table 4, some authors in this study give other arguments for the need for residents' collective voice to be heard [43] and included [51,55]. They claim that residents should even take the initiative in some aspects of housing development [56]. For instance, in the Danish climate policy context, Sørensen and Torfing [55] (p. 13) maintain that

> ... with its emphasis on needs-based problem-solving, knowledge-sharing, joint risk assessment, coordinated and adaptive implementation, and shared ownership of new and bold solutions, co-creation offers a near-perfect strategy for achieving highly ambitious climate mitigation goals.

These arguments suggest that citizenship can easily be applied to the context of housing development and can identify what the concerned researchers perceive as shortcomings in the PPP's ability to engage, or listen to, residents concerning the development of housing in the city's neighbourhood landscape.

3.1.3. Justice

Justice, together with social networks, is by far one of the largest components of the social foundation of housing development (see Table 1). This important result reveals a need for PPPs to better understand how its housing development influences the dynamic of social justice in relation to sustainability. Seventeen (68%) of the articles pertain to justice in one way or another (as in Table 2). From a plethora of critical arguments, three major themes are deduced (as in Table 5). These are related to the elite's power and their documented injustices and role in the (de)stigmatization of so-called deprived neighbourhoods. The relationship between these themes is obvious. Elites use of power can sometimes lead to injustices such as creating rent gaps and stigmatizing neighbourhoods with the intention of emptying them of poor residents (gentrification). What is termed here as social sustainability via eviction.

Table 5. Arguments for the construct of "justice in housing development".

No.	Themes	Sources	Arguments for Justice
1.	Elite power	[40,50–52,60,62,64,65, 67,70,72]	Avoid government (state) steering [52,60,65] Avoid privileging certain sectors, while marginalizing others: social sustainability via eviction [50,62,64,67] Deliberate the fact that joint forces of the elite displace long-time inhabitants [40,70,72] Promote ceding city planning power to citizens [51,70]
2.	Injustice	[51,58,61,63–65,73,74]	Counteract negative effects of gentrification [57] Deliberate conflict resolution in land-use [61,63] Promote revamping distressed neighbourhoods [64] Include all stakeholders in a specific governable context [65] Introduce strong social focus on BID property development [73,74]
3.	Stigmatization	[40,71]	Avoid redevelopment through stigmatization of neighbourhood [40] Be wary of territorial destigmatization regimes [71]

Concerning the elite power theme intrinsic to justice, researchers suggest that four steps can be taken toward justice (as in Table 5). The common denominator for their argumentation is the need for new approaches to avoid an uneven distribution of housing resources [52,60]. To avoid this, la Cour and Andersen [65] suggest a new form of collaboration: metagovernance. They [65] (p. 920) state that

The shift from government to metagovernance ... represents an extraordinarily radical displacement of the contract's form. These new forms of collaboration are bringing about revolutionary changes in the traditional relationship between municipalities and housing associations.

This implies that the PPP shares power with [51,70], and includes the needs of, marginalized residents vis a vis the housing association [50,62,64]. Berglund-Snodgrass et al. [67] (p. 877) even go as far as to argue for the inclusion of social services and marginalized residents:

By primarily organizing settings and knowledge that render familiar to a technocratically governed urban planning, the social services struggle to get recognition in the process or fail to see how their working processes and situated knowledge can be incorporated in the housing provision planning – and are, as a consequence, marginalized in the process.

For instance, the PPP as an elite should not, it is argued by Noring [70] and Andersen et al. [72], be allowed to displace residents. By including the social services in new collaborations such as metagovernance, the elite can be dissuaded to displace. Displacement is, in this reading of the literature, however, a common occurrence.

The theme of injustice is derived from research that specifies different ways to avoid what the authors view as a predominantly unfair housing development. The principles of fairness they suggest in their argumentation, and listed here, ought to be viewed as a form of triage that includes arresting negative neighbourhood effects. This, they claim, can only be done by diluting the current strong focus on Business Improvement Districts (BID) property development with social sensitivity [58,64]. In the Danish context, Richner and Olesen [73] (p. 167) capture this line of argument when they argue that:

> ... the particularities of how the BID model is being translated into the Danish context should not be misread as a case in which the strong Danish social welfarist tradition has mitigated the 'neoliberal aggressiveness' of the BID model.

Thus, and as means of circumventing the ends of neoliberal aggressiveness, BID property development should, according to Candel Candel, Karrbom Gustavsson, and Eriksson [61], also include solutions that satisfies the preferences of all actors, specifically meeting the particular needs of the bureaucratic and political municipality in terms of social equity and political voice and, thus, the general requirements of social sustainability.

Another aspect of (in)justice intrinsic to housing development is the use of the broader phenomenon of stigmatization and responding with the method of destigmatization to redevelop a neighbourhood. Returning once again to the case from Kvillebäcken in Gothenburg, Sweden, we lean on Thörn and Holgersson's [40] (p. 380) illustration of the anatomy of destigmatization and its end product, displacement,

> ... to unravel how the joint forces of the elite (in our case the close cooperation between private real estate owners and the municipality) stigmatizes areas, make the inhabitants invisible and then displace them to favour financial profit.

As a reaction to this unwanted outcome, some researchers suggest that the PPP or other similar collaborations ought to focus instead on a process of destigmatization here defined "as interventions, initiatives, processes or strategies carried out with the intention of reducing, removing, redirecting or remedying the territorial stigmatization of specific places" [71] (p. 1). Schultz Larsen and Delica [71] show, moreover, that this phenomenon is also a wicked problem since it too leads to displacement, and via its Sisyphean character, it "has become a legitimation of the current radical policy measures of demolition, eviction, gentrification and reprivatisation of the stigmatized territories" [71] (p. 17). In sum, the issue of justice, or housing development as fairness, is predominantly a reaction to what scholars perceive as a radical, harmful, aggressive, and socially unsustainable neoliberal housing policy.

3.1.4. Social Networks

As mentioned earlier, social networking is also a big issue that underpins the social foundation of sustainability (Table 1). Seventeen (68%) of the articles that constitute the literature review pertain in one way or another to social networks and their underlying themes (as in Table 6). In the reading, two themes quickly became obvious. The first theme is connectivity and the second collaboration. Connectivity is about the shape or structure of the social network (lines), while collaboration is about which actors are involved (nodes) and how they interact. The link between these two themes and social networks is obvious. If there is a lack of connectivity between stakeholders (developers) and actors (housing associations, social services, and communities as well as residents) concerning recent housing development in the Nordic context, particularly Sweden, then the question that must be answered is if the PPP has a role to play here. Therefore, a substantial number of researchers (as in Table 2) have researched the PPP from these two angles. They have presented several different proposals for the creation of social networks from which are derived eight arguments in Table 6.

Table 6. Arguments for the construct of "social networks in housing development".

No.	Themes	Sources	Arguments for Social Networks
1.	Connectivity	[50,54,57,60,62,68,69,71]	Address policy schizophrenia [71] Consider social structures that encompass most segments of society and avoiding the disconnect between actors [50,54,57,69] Focus on project-bound issue networks, conditioned by local actors [60,62] Promote existing urban governance structures that include key local actors and residents [68,69]
2.	Collaboration	[55,56,59,61,63,66,68,72–74]	Address collective action challenge [63] Construct formal and informal actor-network to mobilize support for urban development [56,73,74] Co-create value via, inter alia, co-management zones [55,59,61] Combine different mode of governing, participation, and coproduction as a counterweight to non-coordinated elite (neoliberal) strategies [66,68,72]

Concerning the connectivity theme, there is an obvious need to reshape the social network in a way that, according to Schultz Larsen and Delica [71] (p. 17), addresses what they term as policy schizophrenia defined here as "fragmentations, splits and contradictions of the current policy regime of housing development". It is argued that the collaborative dimension of social network ought to have both a formal and an informal interaction character [56,73]. A first step in this direction is linked to collaboration and presented by Storbjörk, Hjerpe, and Glaas [74] (p. 582) when they lift several Swedish cases where what they coin the term "developer dialogue", which was applied to encourage public and private actors to "pull together" to mitigate climate change via housing development.

> Malmö, with the district of Västra Hamnen, is often presented as a successful case where developer dialogue facilitated learning and knowledge exchange among property developers and municipal coordinators ... Combining district-level planning with strategies that spur willingness to excel and give credit to those who goes beyond business-as-usual is potentially one way forward here.

Developer dialogue is just one way put forward to scaffold complex stakeholder networks in housing development, particularly when addressing the challenges that face the city and its communities [63]. However, it lacks the ability to include all actors. As a means to the end of widening this collaborative approach, a new point of departure is introduced. This implies co-creating innovative solutions for complex problems (see, for instance, [55,59,61]). This segment of the reviewed research claims that this specific kind of interaction can counteract the negative effects of housing development associated with one-sided neoliberal housing strategies with major legitimacy deficits [66,68,72].

In order to achieve this, the PPP, according to researchers such as Olsson et al. [50], Borgström [54], Glaas et al. [57], and Gohari et al. [69], must permeate and connect with all of society's social strata. Taking the Swedish context as an example, the reason why this is necessary becomes obvious. When studying housing development in Stockholm, Borgström [54] (p. 472) says that

> *The disconnect we found was a bit surprising, given the long-term Swedish tradition of involving and interacting with civic associations, which can be interpreted as good grounds for trust, communication and collaboration.*

This finding indicates that most PPPs in Sweden (like elsewhere) do not just follow neoliberal strategies but may have problems with engendering trust and maintaining lines of communication with residents. As a counterweight to the absence of local actors in PPP networks, some Swedish researchers argue for the implementation of what Elander and Gustavsson [60] have coined "project-bound issue networks". This ought to include, besides the "usual suspects" of the elite, local actors (see also [62]).

> *Viewing social inclusion in this broader context, individuals could increase their social capital and thereby make themselves better able to participate in local planning and politics, perhaps even by acting as "everyday makers".* [60] (p. 1095)

This coincides with the aspirations of another cohort of the small research community in focus in our study. They argue for the promotion of the idea of social inclusion and participation in what they term as the existing urban governance structure [68,69]. This may be a solution to the policy schizophrenia referred to by Schultz Larsen and Delica [71].

3.2. Ecologically Sustainable Housing Development

Regarding the ecological foundation of housing development, and guided by the reading of the latest research, concerning PPP in the Nordic countries, two issues and six themes have surfaced. Guided by the DE-model [5], these issues and themes together outline the potential role of PPP in terms of its ability to tackle the ecological sustainability challenges the city and present as well as future generation communities are facing. That is, ways in which PPP ought to tackle climate change and the combined issue of contributing to biodiversity, on the one hand, and minimizing land conversion and preserving biodiversity, on the other. Derived from Stopper et al. [76] version of the DE model climate change is defined as supply chain management, reduction of CO_2 emissions, energy consumption reduction, increased energy efficiency, and renewable energy use such as biofuel [76]. While biodiversity and land-conversion are defined as the conservation of regional species and use of raw materials produced by organic farming, effective use of old industrial sites, and laying out greens space, respectively [76]. In the following section, each issue and inherent theme mirrors the potential of PPP to become more sustainable with regards to these two planetary boundaries (PBs) of ecological sustainability [4].

Aligned to the social side of the sustainability coin, one initial tangible result is that in its present role, PPP does not tackle the full spectrum of ecological issues as PBs in the DE model [4,5]. From the reading of the research on PPP in housing development, it is apparent that neither the PBs of air pollution, chemical pollution, ozone layer depletion, ocean acidification, freshwater withdrawals nor nitrogen and phosphorus loading are considered. The remaining two issues have undergone a similar softening up process as the social issues. That is, the same modus operandi is applied to synthesize Raworth's [5] broader issues with a deeper critical perspective on the PPP provided by the authors of the reviewed research. In this way, the analysis will also reveal the ecological potential of PPP in sustainable housing development.

3.2.1. Climate Change

To reiterate, cities today are at risk of facing the increasing negative consequences of climate change while they themselves are responsible for 75% of the world's emissions with regards to energy use [1]. Nine (36%) of the reviewed articles touch on the issue of climate change (Table 1). In the reviewed research, three themes and ten arguments intrinsic to climate change were identified. These are based on how PPP in housing ought to tackle the multitudes of challenges concerning tackling climate change via participation, mitigation, and adaption (as in Table 7).

Table 7. Arguments for the construct of "climate change in housing development".

No.	Themes	Sources	Arguments for Climate Change
1.	Participation	[50,55,66,68,70,74]	Climate change tackled through co-creation, participation, and co-production [50,55,68–70] Promote local Governance [66] Sharp goals in public-private interplay [74]
2.	Mitigation	[55,57,60,66,69,74]	Energy efficiency, energy positive, and fossil free power [55,57,60,66,69] Challenge mainstream building practices [66] Consumption and transport behaviour [55,74] Visualisation and measurements [57]
3.	Adaptation	[50,57,74]	Ecosystem services [50] Mitigation of flooding [50,57,74] Adaptation to heath stress [57]

As noted above, several researchers identified the need for PPP to get involved in different forms of participation strategies to tackle the complex issue of climate change in the urban environment (see also Table 4). When it comes to ambitious climate goals, where the PPP and citizens must become involved, co-creation is considered a "near-perfect" strategy [55]. Even if citizen participation is often marginal in projects tackling climate change, it is nevertheless essential in the attainment of tangible results [50,69]. It is believed that if a platform for participation is created, where both stakeholders and citizens can express their opinions and ambitions directly, the process will be both effective and democratic [69].

Other authors are more reserved claiming that a participatory strategy offers a promise but not a perfect solution to climate change since there are several barriers that need to be addressed [68]. For instance, and flipping the participation coin, in collaborations between public and private companies, private companies tend to downplay high climate goals [74]. To overcome this barrier, Storbjörk et al. [74] (p. 582) suggest

> ...the steering strategies used by public actors to secure the realization of key public goals such as climate change in urban development needs to be refined and sharpened, particularly at the stage of sustaining commitments and securing formal agreements.

The second theme that emerged on how PPP can tackle climate change is through mitigation. Examples of technology application to support mitigation strategies include energy efficiency [60], reductions in district heating and the proliferation of windfarms [55], maintaining high requirements for energy [69], guaranteeing fossil free power utility [57], and following specific technical requirements and standards [66]. To support this kind of technology-transition, Smedby and Quitzau [66] (p. 332) suggest local government have an important role to play, for instance:

> Local governments proactively engage in a balancing act aiming at integrating radical innovations and mainstream construction practices to foster the transition towards sustainable socio-technical systems.

Some researchers bring attention to the problematic of technological solutions promised by the "smart city" approach, particularly when technical, economic, and political goals are frequently prioritized over social and environmental goals. However, another solution to unsustainable development is, to reiterate, to include citizens, communities, local associations, as well as concerned PPP stakeholders in the smart city approach [69].

For cities and PPPs to keep the global temperature well below an increase of 2 °C (agreed in the Paris Agreement) and to mitigate the worst effects of climate change, challenges such as changing citizens' consumption and transport behaviour need to be addressed [55]. Some suggestions concern changing mobility patterns by reducing parking lots and introducing carpools, avoiding floor heating [74], and reducing emissions in construction [50]. In addition, the measurement and visualization of climate change effects need to be combined with clear targets and a systemic understanding if urban climate transition is to be achieved [57].

To adapt to the effects of climate change, and to prevent extreme weather events, some researchers suggest ecosystem services (ES) as a strategy for PPP [50]. In one project, adaptation strategies such as storm water mitigation through ponds or green areas were part of a vision to mitigate the negative effects of climate change [74], and in another project, progress was made in adapting buildings for heat stress [57]. Once again, the local perspective is identified as being important to achieve a just adaptation according to Olsson et al. [50] (see Table 7 above). Olsson et al argue [50] (p. 312)

> ... that there is a need to measure and map the ES provision at the neighbourhood level in relation to the needs of divergent stakeholder groups, understanding the trade-offs between local and city needs.

Researchers suggest that for PPP to tackle these complex challenges of climate change, both mitigation and adaptation strategies need to be addressed simultaneously. Technological solutions will contribute, if, and only if, they are not prioritized over socio-ecological goals and targets. For PPP to adopt these ecological strategies, stakeholders will need to engage in co-creation and participatory strategies with residents and other local stakeholders.

3.2.2. Biodiversity Loss and Land Conversion

Biodiversity and land conversion are two PB's that are central to housing development and have transcended beyond just being a safe space for humanity [4]. For instance, land use policy can impact housing provision through incentives and restrictions [78]. In this case, the two PB's are combined since they are innately interconnected. This implies that first order effects in land conversion might cause second order effects for biodiversity and vice versa [79,80]. By far the smallest issue, touched on by a mere five (20%) of the articles in the review, biodiversity and land conversion, has three themes. These are: anthropocentrism, collaboration, and inaction and divestment. In all, and because they are a criticism of PPP involved in housing development, these themes are deduced from five arguments put forward by the authors in the review (as in Table 8).

Table 8. Arguments for the construct of "Biodiversity and land conversion in housing development".

No.	Themes	Sources	Argument for Biodiversity and Land Conversion
1.	Anthropocentrism	[50,56,59]	Residents need for green space [50,56,59]
2.	Collaboration	[56,59]	Stakeholder involvement important [56] Co-management in urban forestry [59]
3.	Inaction and divestment	[51,57,59]	Biodiversity and land use are subjects of inaction [57,59] Avoid divestment of land by municipalities [51]

Concerning anthropocentric needs, Olsson et al. [50] recognize a need among residents for green space in, or near, their neighbourhoods. Here, access to green space is underpinned by both social and ecological arguments (see also Table 3). Nevertheless, with regards to urban farming and food production, green spaces such as community gardens only have a marginal contribution to sustainable development in the city in terms of instrumental value [56]. Nevertheless, there is still support for the idea of creating and developing community gardens to further contribute to sustainable development, Bonow and Normark [56] (p. 515) suggest

> ...municipalities and housing companies should also focus on knowledge support, as well as providing some physical prerequisites for growing (access to water, etc.).

In addition, for community gardens to become more sustainable with regards to food production, Bonow and Normark [56] suggest the involvement of stakeholders from NGO's, the municipality, and housing companies to facilitate the processes further. This is evidence that there is a potential role for PPP to play in this context. Similarly, Fors et al. [59] (p. 54) discover once again the importance of collaboration (see Table 6), and

> ...emphasizes the need for continuous municipality-resident communication, including municipal guidance, inspiration and control.

However, community gardens [56], and public woodland also have a well-documented recreational value for both present as well as future generation residents [59]. Therefore, green areas have, as noted above (Table 3), an even greater significance for social sustainability.

When it comes to biodiversity and land conversion, human intervention in nature is in focus, while the enrichment and preservation of species (biodiversity) and nature in human "space" are less common [59]. In urban climate transition, it is common with inaction with regards to biodiversity, forestry, and agriculture while most focus tends to go to energy solutions and activities [57]. Fors et al. [59] suggest that this aspect needs further research to find solutions that benefit both biodiversity and the urban environment. A case from Finland shows that when a municipality sells public land to housing developers, it loses control over both its use in terms of urban farming and recreation (biodiversity) as well as housing prices. Hyötyläinen and Haila [51] (p. 144) were critical of this kind of development in a project in Finland; Helsinki

> ... Eiranranta was an experiment by the City to test the upper end of the housing market we can just hope this experiment does not lead to more selling off of public land.

In conclusion, human needs are today the priority and the guiding principle for PPPs when converting land for housing development, while biodiversity and land conversion are not that prioritized. In order to create a sustainable city, all of these three aspects, human needs, biodiversity and land use, will need to be prioritized by the PPP in an equal manner. Municipalities and the PPP need, thus, to avoid selling land and falling into inaction with regards to biodiversity and the segregation of whole communities and their neighbourhoods negating the possibility of creating a harmonious urban environment. Housing companies, municipalities, NGOs, housing associations, and residents are recommended to collaborate and participate in safeguarding nature and the city's ecosystems.

4. Discussion

Given the fact that the complexity of sustainability in the urban environment often exceeds the capacity of an individual organization [6,7], the potential role of PPP in sustainable urban development and renewal was investigated and found to be crucial. Applying the DE model [5], the systematic literature review shows how housing development can become more sustainable if certain identified issues and themes are brought to the attention of, and internalized by, the PPP in the future.

However, to meet the needs of local housing development in the Nordic context, the DE model needs further revision [5]. Scaling down from PBs and SDGs to local problems such as inequalities is, however, not a trivial task. To this end highlighting features of " ... the harmonious evolution of civil society, fostering an environment that encourages social integration, with improvements in the quality of life for all segments of populations" [81] (p. 19) was essential.

Moreover, consideration was taken of the fact that the DE model focuses on countries irrespective of their political systems. As it stands, the ecological foundation of the DE model disregards differences between places. However, the Nordic-countries are representative democracies characterized by parliamentarism and the condition of moderate scarcity [82], and with increasing spatial segregation in mind, some issues such as gender equality, political voice, and education are aspects of social sustainability that still remain problematic in a democracy. Likewise, with climate change in mind, an open mind is essential concerning the issues of food, water, and energy, which do not appear to be a major concern for social sustainability in the Nordic countries. Similarly, a reflection on which aspects of ecological sustainability PPP can and cannot influence needs to be undertaken. For instance, ocean acidification or phosphorus loading seem to be beyond its reach. For this reason, a holistic and systemic point of view was applied [42,83]. Thus, the DE model and all its issues are applied and left open for discussion.

4.1. Participation and Collaboration

First and foremost, the authors find that broader participation and more inclusive collaboration in PPP is crucial if cities in the Nordic countries are to move in the direction of sustainable housing development. These two themes constitute a common denominator for achieving both types of sustainability. They also constitute a key argument for reforming PPP. For instance, PPP needs to reconsider the importance of residents' preferences by promoting their participation in the development and renewal of urban areas. Furthermore, PPP needs also to move beyond present collaboration to new forms such as co-creation with residents and other community actors. From the point of view of the reviewed research, this would counteract and circumvent the current negative effects of neoliberal housing strategies. It is also a way to mitigate climate change and spatial segregation as well as contribute to more access and biodiversity in the city. Mang and Reed [84] corroborate this view when they too show that a participatory design is an effective and systemic strategy to engage residents and maintain trajectory toward a sustainable and regenerative society.

4.2. Justice and Social Networks

Justice is the first big issue in the review (see Table 2). Winston and Eastaway [85] corroborate this. They say that social sustainability is about guaranteeing equal opportunities in new housing development [85]. This research shows that justice as fairness can be expressed in the micro context of the neighbourhood in terms of adequate domestic living space, affordable housing, and resistance to crime and in the macro context of the city in terms of reduced social spatial segregation [86]. It is clear from the review that the research community, governments, and their national policies, as well as international treaties need to deliberate the power that the joint forces of PPP are wielding. The question at issue, with the future city in mind, is if these stakeholders are willing to cede their power to better serve the people. Hence, if the PPP is to take heed to the issue of justice, it needs to understand, with the case of Kvillebäcken addressed by Thörn and Holgersson [40], how it can avoid destigmatization processes. In essence, PPP has the potential to buttress the city's need to live in tune with its society and environment. Ultimately, a greater understanding of the power dynamic of PPP in sustainable housing development is needed to achieve this.

The importance of stretching and deepening participation and collaboration is most evident in the issue of social networks, which is also touched on by many researchers in the reviewed articles (see Table 1). The review reveals that there is a need for cultivating networks made up primarily of people, non-profit civic organizations, and PPP [76]. For this kind of public-private-people partnership (PPPP) to come about researchers say that current structures need to transcend social strata, thus, bridging the distance between diverse groups in society [54,57,69]. Another argument for bringing different actors together under the umbrella of PPPP is that it encourages dialogue [74].

One major discrepancy was found between the issues intrinsic to the DE-model and the reviewed articles (see Figure 3), on the one hand, and the UN's universal SDG 11 and its subgoals, on the other. In the review the spotlight was on social networks and justice. However, in the subgoals of SDG 11, neither social network nor justice are addressed to any greater length (see Table 1). This is an interesting finding since most researchers in the review regard these issues as the main weaknesses of the housing development PPP. This is unsurprising since the authors are not applying all SDGs to the case of PPP in housing development. This implies that even if PPP adhered to the goals of SDG 11, it would still need to address all the SDGs and PBs to tackle the sustainability challenges that future cities are facing.

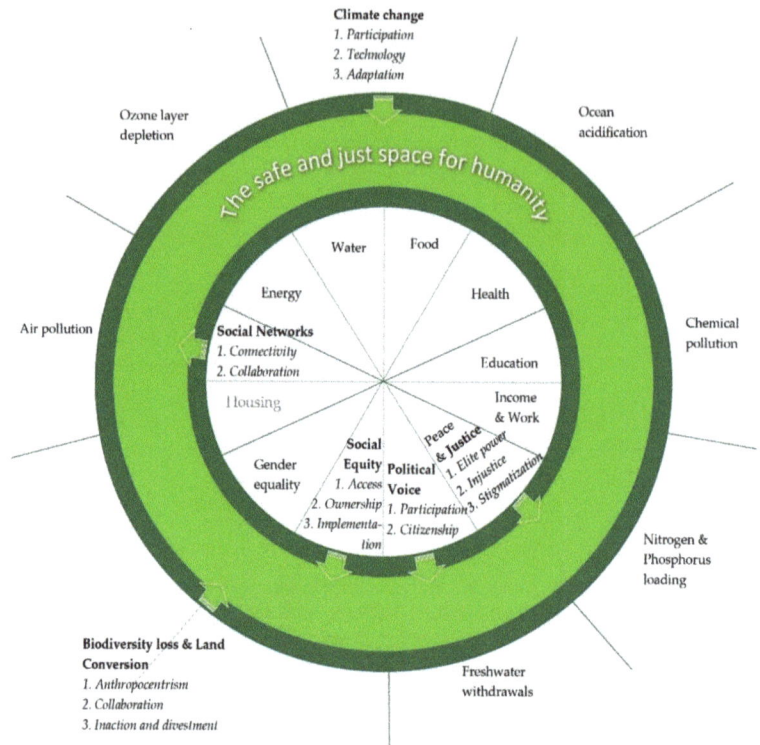

Figure 3. The potential of PPP with regard to the localized DE-model.

4.3. Superimposing Local Context Issssues on the DE Model

The method applied here shows that there are a number of similarities and differences between the character of the same issues that connect the global model of sustainability to the harsh reality of the local context now superimposed on Figure 3. Social equity, justice, social networks, climate change, and biodiversity loss and land conversion are good examples of issues that are heavily criticized and, thus, change character by expanding into multiple themes at the local level of the PPP, while political voice is a good example of an issue that does not change character after transcending the global-local divide. Nevertheless, this implies only that local researchers amplify DE model theorists' calls for more participation and citizenship, which the authors of this review maintain that PPP has the potential to achieve. The connection that bridges the PPP-sustainability divide is still weak and needs to be strengthened in accordance with the review's results.

Concerning the PPP-sustainability gap, it is obvious that issues that underpin the future city's social foundation such as gender equality, health, education, income and work, food, water, and energy need to be addressed more in-depth in housing development in the Nordic countries. Consideration also needs to be taken of the gap in housing development's ability to tackle issues related to the future city's ecological ceiling such as freshwater withdrawals, nitrogen and phosphorous loading, chemical pollution, ocean acidification, ozone layer depletion, and air pollution (Figure 3).

4.4. A Holistic and Systematic Approach

A major concern for researchers is that the gap identified in the modified DE-model (Figure 3) cannot inform about those issues the review did not explicitly mention such as food (apart from community gardens) and ocean acidification. Noteworthy is that energy and water seem in the Nordic countries to be an ecological problem not a social one. What can be said, however, is that there is a lack of research in the reviewed literature on how PPP can tackle these issues.

For the city to become sustainable, PPP will need to adopt a holistic point of view and address more sustainability issues. For instance, one of the issues that is missing from the repertoire of PPP is pollution. Lowering chemical pollution in construction through reusable and recycled materials [87] or densifying the city to mitigate air pollution [88] are two strategies that PPP has the capacity to influence or adopt. Using the same line of holistic reasoning concerning health, PPP housing development needs to guarantee access to public green space, even if it is concentrated in areas of dereliction [85].

5. Conclusions

First, if the collaborations that constitute PPP are to be used to develop cities, their responsibility must go far beyond developing housing. For PPP to contribute to future sustainable urban development and renewal, they will need to address both social and ecological issues in a more systematic, participatory, and collaborative manner. Adding a fourth P, people, to PPP might be a first step in the right direction for this to transpire.

The second concerns the application of the DE model. This review article shows that the DE model can be used in a normative sense, that is, to test the scope and depth of local collaborations such as PPPs and reflect on international treaties such as SDGs. The application of the DE model in this article is a proof of concept that reveals both the shortcomings of PPP and SDG 11. The revised DE model transcends beyond the notions of sustainable development expressed in SDGs to create a more social and ecological sustainable city. It can also be applied to various forms of collaboration with a focus on any of its DE model's issues.

Third, the DE model reveals the need for a better connection between global sustainability and the PPP's potential to address certain issues such as justice and contribute to the sustainability of the future city, appeal to its communities, and move beyond the limitations of SDG 11. The DE model also reveals a gap in terms of the issues not touched on in the reviewed research that must be addressed after the mentioned connection is strengthened.

Fourth, and based on the results, it was found that only seven of the 21 issues touched on by the DE model are covered by the reviewed research. Another result reveals that PPPs in this study are always being criticized and mostly for their lack of social sustainability. This undermines the social foundation of the future city and its communities. What is also striking is the fact that the two main issues, justice and social networks, touched on by most researchers in the review are those not covered by SDG 11. In essence, a new gap in knowledge becomes apparent in a comparison of the DE-model with the subgoals of SDG 11. Consequently, the PPP has not yet tackled the full spectrum of the social foundation of sustainability, as depicted by Raworth [5] and Stopper et al. [76] in the DE model. In fact, as it stands, it seems to ignore gender equality, and neither does it appear to promote education, guarantee income, and work nor cater for the health of residents.

Fifth, based on suggestions from researchers in the review, new planning perspectives that include residents' and developers' preferences ought to be adopted by the stakeholders that constitute the PPP if future housing development is to be built upon a solid foundation of social equity. They imply that this will not be possible if the PPP continues to rely on current neoliberal justifications. To reiterate, these arguments also suggest that citizenship can easily be applied to the context of housing development. Moreover, a focus on citizenship can identify what the concerned researchers perceive as shortcomings in the PPP's ability to engage, or listen to, residents concerning the development of housing in the city's neighbourhood landscape. In sum, the issue of justice, or housing development as fairness, is predominantly a reaction to what scholars perceive as a radical, harmful, aggressive, and socially unsustainable neoliberal housing policy.

Furthermore, another cohort of the small research community expressed the need for more robust social networks that coincides with the aspirations of in focus in the study. They argue for the promotion of the idea of social inclusion and participation in what they term as the existing urban governance structure. As mentioned earlier, this may be a solution to the policy schizophrenia referred to by Schultz Larsen and Delica [71]. Researchers also suggest that for PPP to tackle the complex challenges of climate change, both mitigation and adaptation strategies need to be addressed simultaneously. Human needs are today the priority and the guiding principle for PPPs when tackling issues of justice, climate change, and converting land for housing development. Finally, and in order to create a sustainable city, all of these issues will need to be prioritized by the PPP in an equal manner.

6. Limitations and Future Research

Finally, and highlighting some current limitations of this study, the article does not address the positive aspects of PPP. Instead, focus was on deriving its future potential to achieve sustainability in the city from critical accounts in the literature. Therefore, the authors only focused on criticism of current PPP policy goals in the Nordic countries. The analysis is also limited to seven issues and can only scratch the surface concerning the significance of the remaining 14 issues. For instance, just because current research (2015–2021) does not mention education, pollution, and water, it does not imply that PPPs are avoiding these issues.

Our article has some implications that need to be addressed by future research. Firstly, the research community needs to know if PPP has the necessary and sufficient institutions to go from potentiality to actuality and from being an isolated problem-solver to becoming a systematic and inclusive player, an avant-garde, in tackling urban unsustainability. Secondly, it is important to determine what facilitates or hinders the movement of PPP towards sustainability. In essence, what will it take for the stakeholders that constitute PPP to get on-board and engage in the process of enabling a transition toward a more sustainable future city? Finally, future research is recommended to find more ways to apply the DE model to the varying contexts of the city and support the transition toward a sustainable urban development.

Author Contributions: Both authors have made a substantial contribution to the conception and design of the work as well as the acquisition, analysis, and interpretation of data for the work. We have also been involved in drafting and reviewing the work critically. We give our final approval of this version to be published. We agree to be accountable for all aspects of the work in ensuring that questions related to the accuracy or integrity of any part of the work are appropriately investigated and resolved. In more detail: conceptualization T.F. and J.M., methodology and software, J.M.; writing—original draft preparation, supervision and validation T.F., formal analysis, review and editing T.F. and J.M. All authors have read and agreed to the published version of the manuscript.

Funding: This work has been carried out under the auspices of the industrial post-graduate school Future Proof Cities (grant number 2019-0129), which is financed by the Knowledge Foundation (KK-stiftelsen). We kindly thank the funding bodies for their financial support.

Institutional Review Board Statement: Not applicable.

Informed Consent Statement: Not applicable.

Acknowledgments: We would like to thank Rana Mostaghel, Lars Lindbergh, Carin Nordström, and the DEM-group at Mälardalen University for sharing their expertise and knowledge and commenting on drafts of the paper.

Conflicts of Interest: The authors declare no conflict of interest.

References

1. Bai, X.; Dawson, R.J.; Ürge-Vorsatz, D.; Delgado, G.C.; Barau, A.S.; Dhakal, S.; Roberts, D. Six research priorities for cities and climate change. *Nature* **2018**, *555*, 23–25. [CrossRef]
2. Tanguay, G.A.; Rajaonson, J.; Lefebvre, J.-F.; Lanoie, P. Measuring the sustainability of cities: An analysis of the use of local indicators. *Ecol. Indic.* **2010**, *10*, 407–418. [CrossRef]
3. Chan, P. Assessing Sustainability of the Capital and Emerging Secondary Cities of Cambodia Based on the 2018 Commune Database. *Data* **2020**, *5*, 79. [CrossRef]
4. Steffen, W.; Richardson, K.; Rockstrom, J.; Cornell, S.E.; Fetzer, I.; Bennett, E.M.; Biggs, R.; Carpenter, S.R.; de Vries, W.; de Wit, C.A.; et al. Sustainability. Planetary boundaries: Guiding human development on a changing planet. *Science* **2015**, *347*, 1259855. [CrossRef] [PubMed]
5. Raworth, K. A Doughnut for the Anthropocene: Humanity's compass in the 21st century. *Lancet Planet. Health* **2017**, *1*, e48–e49. [CrossRef]
6. Fobbe, L. Analysing Organisational Collaboration Practices for Sustainability. *Sustainability* **2020**, *12*, 2466. [CrossRef]
7. Lozano, R. Collaboration as a pathway for sustainability. *Sustain. Dev.* **2007**, *15*, 370–381. [CrossRef]
8. Pero, M.; Moretto, A.; Bottani, E.; Bigliardi, B. Environmental collaboration for sustainability in the construction industry: An exploratory study in Italy. *Sustainability (Switzerland)* **2017**, *9*, 125. [CrossRef]
9. Loorbach, D.; Wijsman, K. Business transition management: Exploring a new role for business in sustainability transitions. *J. Clean. Prod.* **2013**, *45*, 20–28. [CrossRef]
10. Rockström, J.; Steffen, W.; Noone, K.; Persson, Å.; Chapin III, F.S.; Lambin, E.; Lenton, T.M.; Scheffer, M.; Folke, C.; Schellnhuber, H.J. Planetary boundaries: Exploring the safe operating space for humanity. *Ecol. Soc.* **2009**, *14*. [CrossRef]
11. Hoornweg, D.; Hosseini, M.; Kennedy, C.; Behdadi, A. An urban approach to planetary boundaries. *Ambio* **2016**, *45*, 567–580. [CrossRef]
12. Boffey, D. *Amsterdam to Embrace'Doughnut'Model to Mend Post-Coronavirus Economy*; The Guardian: London, UK, 2020; Volume 8.
13. Luukkanen, J.; Vehmas, J.; Kaivo-Oja, J. Quantification of Doughnut Economy with the Sustainability Window Method: Analysis of Development in Thailand. *Sustainability* **2021**, *13*, 847. [CrossRef]
14. Roy, A.; Basu, A.; Dong, X. Achieving Socioeconomic Development Fuelled by Globalization: An Analysis of 146 Countries. *Sustainability* **2021**, *13*, 4913. [CrossRef]
15. Saunders, A.; Luukkanen, J. Sustainable development in Cuba assessed with sustainability window and doughnut economy approaches. *Int. J. Sustain. Dev. World Ecol.* **2021**, 1–11. [CrossRef]
16. Osei-Kyei, R.; Chan, A.P.C. Review of studies on the Critical Success Factors for Public–Private Partnership (PPP) projects from 1990 to 2013. *Int. J. Proj. Manag.* **2015**, *33*, 1335–1346. [CrossRef]
17. Hodge, G.A.; Greve, C. On Public–Private Partnership Performance. *Public Works Manag. Policy* **2016**, *22*, 55–78. [CrossRef]
18. OECD. *Public-Private Partnerships: In Pursuit of Risk Sharing and Value for Money*; Organisation for Economic Co-operation and Development (OECD: Paris, France, 2008.
19. Osei-Kyei, R.; Chan, A.P. Perceptions of stakeholders on the critical success factors for operational management of public-private partnership projects. *Facilities* **2017**, *35*, 21–38. [CrossRef]
20. Opawole, A.; Jagboro, G.O.; Kajimo-Shakantu, K.; Olojede, B.O. Critical performance factors of public sector organizations in concession-based public-private partnership projects. *Prop. Manag.* **2019**, *37*, 17–37. [CrossRef]
21. Ameyaw, E.E.; Chan, A.P. Critical success factors for public-private partnership in water supply projects. *Facilities* **2016**, *34*, 124–160. [CrossRef]
22. Cheung, E.; Chan, A.P.; Kajewski, S. Factors contributing to successful public private partnership projects: Comparing Hong Kong with Australia and the United Kingdom. *J. Facil. Manag.* **2012**, *10*, 45–58. [CrossRef]
23. Parashar, D. The Government's role in private partnerships for urban poor housing in India. *Int. J. Hous. Markarkets Anal.* **2014**, *7*, 524–538. [CrossRef]
24. Ameyaw, E.E.; Chan, A.P. Evaluating key risk factors for PPP water projects in Ghana: A Delphi study. *J. Facil. Manag.* **2015**, *13*, 133–155. [CrossRef]
25. Babatunde, S.O.; Opawole, A.; Akinsiku, O.E. Critical success factors in public-private partnership (PPP) on infrastructure delivery in Nigeria. *J. Facil. Manag.* **2012**, *10*, 212–225. [CrossRef]
26. Alteneiji, K.; Alkass, S.; Abu Dabous, S. Critical success factors for public–private partnerships in affordable housing in the United Arab Emirates. *Int. J. Hous. Mark. Anal.* **2019**, *13*, 753–768. [CrossRef]
27. Kavishe, N.; Chileshe, N. Critical success factors in public-private partnerships (PPPs) on affordable housing schemes delivery in Tanzania. *J. Facil. Manag.* **2019**, *17*, 188–207. [CrossRef]
28. Trangkanont, S.; Charoenngam, C. Private partner's risk response in PPP low-cost housing projects. *Prop. Manag.* **2014**, *32*, 67–94. [CrossRef]
29. Jegede, F.O.; Adewale, B.A.; Jesutofunmi, A.A.; Loved, K.S. Assessment of Residential Satisfaction for Sustainability in Public-Private Partnerships (PPPs) Housing Estates in Lagos State, Nigeria. In Proceedings of the IOP Conference Series: Earth and Environmental Science, Covenant University, Ota, Nigeria, 23–20 June 2020.
30. Della Spina, L.; Calabrò, F.; Rugolo, A. Social housing: An appraisal model of the economic benefits in Urban regeneration programs. *Sustainability (Switzerland)* **2020**, *12*, 609. [CrossRef]

31. Kavishe, N.; Jefferson, I.; Chileshe, N. Evaluating issues and outcomes associated with public–private partnership housing project delivery: Tanzanian practitioners' preliminary observations. *Int. J. Constr. Manag.* **2019**, *19*, 354–369. [CrossRef]
32. Shi, J.; Duan, K.; Wen, S.; Zhang, R. Investment valuation model of public rental housing PPP project for private sector: A real option perspective. *Sustainability (Switzerland)* **2019**, *11*, 1857. [CrossRef]
33. Lohri, C.R.; Camenzind, E.J.; Zurbrügg, C. Financial sustainability in municipal solid waste management—Costs and revenues in Bahir Dar, Ethiopia. *Waste Manag.* **2014**, *34*, 542–552. [CrossRef]
34. Nizkorodov, E. Evaluating risk allocation and project impacts of sustainability-oriented water public–private partnerships in Southern California: A comparative case analysis. *World Dev.* **2021**, *140*. [CrossRef]
35. Kivilä, J.; Martinsuo, M.; Vuorinen, L. Sustainable project management through project control in infrastructure projects. *Int. J. Proj. Manag.* **2017**, *35*, 1167–1183. [CrossRef]
36. Fraser, J.C. Beyond Gentrification: Mobilizing Communities and Claiming Space. *Urban Geogr.* **2004**, *25*, 437–457. [CrossRef]
37. Siemiatycki, M. Urban Transportation Public–Private Partnerships: Drivers of Uneven Development? *Environ. Plan. A Econ. Space* **2011**, *43*, 1707–1722. [CrossRef]
38. Smith, S.L. Devising environment and sustainable development indicators for Canada. *Corp. Environ. Strategy* **2002**, *9*, 305–310. [CrossRef]
39. Polk, M. Institutional capacity-building in urban planning and policy-making for sustainable development: Success or failure? *Plan. Pract. Res.* **2011**, *26*, 185–206. [CrossRef]
40. Thörn, C.; Holgersson, H. Revisiting the urban frontier through the case of New Kvillebäcken, Gothenburg. *City* **2016**, *20*, 663–684. [CrossRef]
41. Gray, B.; Purdy, J. Collaborating for Our Future: Multistakeholder Partnerships for Solving Complex Problems. Oxford University Press: Oxford, UK, 2018.
42. Hagbert, P.; Malmqvist, T. Actors in transition: Shifting roles in Swedish sustainable housing development. *J. Hous. Built Environ.* **2019**, *34*, 697–714. [CrossRef]
43. Xiao, Y.; Watson, M. Guidance on Conducting a Systematic Literature Review. *J. Plann. Educ. Res.* **2019**, *39*, 93–112. [CrossRef]
44. Moher, D.; Shamseer, L.; Clarke, M.; Ghersi, D.; Liberati, A.; Petticrew, M.; Shekelle, P.; Stewart, L.A. Preferred reporting items for systematic review and meta-analysis protocols (PRISMA-P) 2015 statement. *Syst. Rev.* **2015**, *4*, 1. [CrossRef] [PubMed]
45. Malek, J.A.; Lim, S.B.; Yigitcanlar, T. Social Inclusion Indicators for Building Citizen-Centric Smart Cities: A Systematic Literature Review. *Sustainability* **2021**, *13*, 376. [CrossRef]
46. Liberati, A.; Altman, D.G.; Tetzlaff, J.; Mulrow, C.; Gøtzsche, P.C.; Ioannidis, J.P.A.; Clarke, M.; Devereaux, P.J.; Kleijnen, J.; Moher, D. The PRISMA statement for reporting systematic reviews and meta-analyses of studies that evaluate health care interventions: Explanation and elaboration. *J. Clin. Epidemiol.* **2009**, *62*, e1–e34. [CrossRef]
47. Grundström, K.; Molina, I. From Folkhem to lifestyle housing in Sweden: Segregation and urban form, 1930s–2010s. *Int. J. Hous. Policy* **2016**, *16*, 316–336. [CrossRef]
48. Baeten, G.; Westin, S.; Pull, E.; Molina, I. Pressure and violence: Housing renovation and displacement in Sweden. *Environ. Plan. A Econ. Space* **2017**, *49*, 631–651. [CrossRef]
49. Ouzzani, M.; Hammady, H.; Fedorowicz, Z.; Elmagarmid, A. Rayyan—A web and mobile app for systematic reviews. *Syst. Rev.* **2016**, *5*, 1–10. [CrossRef]
50. Olsson, J.A.; Brunner, J.; Nordin, A.; Hanson, H.I. A just urban ecosystem service governance at the neighbourhood level-perspectives from Sofielund, Malmö, Sweden. *Environ. Sci. Policy* **2020**, *112*, 305–313. [CrossRef]
51. Hyötyläinen, M.; Haila, A. Entrepreneurial public real estate policy: The case of Eiranranta, Helsinki. *Geoforum* **2018**, *89*, 137–144. [CrossRef]
52. Noring, L.; Struthers, D.; Grydehøj, A. Governing and financing affordable housing at the intersection of the market and the state: Denmark's private non-profit housing system. *Urban Res. Pract.* **2020**, 1–17. [CrossRef]
53. Braun, V.; Clarke, V. Using thematic analysis in psychology. *Qual. Res. Psychol.* **2006**, *3*, 77–101. [CrossRef]
54. Borgström, S. Balancing diversity and connectivity in multi-level governance settings for urban transformative capacity. *Ambio* **2019**, *48*, 463–477. [CrossRef]
55. Sørensen, E.; Torfing, J. Co-creating ambitious climate change mitigation goals: The Copenhagen experience. *Regul. Gov.* **2020**. [CrossRef]
56. Bonow, M.; Normark, M. Community gardening in Stockholm: Participation, driving forces and the role of the municipality. *Renew. Agric. Food Syst.* **2018**, *33*, 503–517. [CrossRef]
57. Glaas, E.; Hjerpe, M.; Storbjörk, S.; Neset, T.-S.; Bohman, A.; Muthumanickam, P.; Johansson, J. Developing transformative capacity through systematic assessments and visualization of urban climate transitions. *Ambio* **2019**, *48*, 515–528. [CrossRef] [PubMed]
58. Lidegaard, C.; Nuccio, M.; Bille, T. Fostering and planning urban regeneration: The governance of cultural districts in Copenhagen. *Eur. Plann. Stud.* **2018**, *26*, 1–19. [CrossRef]
59. Fors, H.; Nielsen, A.B.; van den Bosch, C.C.K.; Jansson, M. From borders to ecotones–Private-public co-management of urban woodland edges bordering private housing. *Urban For. Urban Green.* **2018**, *30*, 46–55. [CrossRef]
60. Elander, I.; Gustavsson, E. From policy community to issue networks: Implementing social sustainability in a Swedish urban development programme. *Environ. Plan. C Politics Space* **2019**, *37*, 1082–1101. [CrossRef]

61. Candel, M.; Karrbom Gustavsson, T.; Eriksson, P.-E. Front-end value co-creation in housing development projects. *Constr. Manage. Econ.* **2021**, *39*, 245–260. [CrossRef]
62. Hermelin, B.; Jonsson, R. Governance of waterfront regeneration projects: Experiences from two second-tier cities in Sweden. *Int. J. Urban Reg. Res.* **2021**, *45*, 266–281.
63. Puustinen, T.L.M.; Viitanen, K.J. Infill development on collectively owned residential properties: Understanding the decision-making process–Case studies in Helsinki. *Hous. Theory Soc.* **2015**, *32*, 472–498. [CrossRef]
64. Valli, C.; Hammami, F. Introducing Business Improvement Districts (BIDs) in Sweden: A social justice appraisal. *Eur. Urban Reg. Stud.* **2021**, *28*, 155–172. [CrossRef]
65. la Cour, A.; Andersen, N.A. Metagovernance as strategic supervision. *Public Perform. Manag. Rev.* **2016**, *39*, 905–925. [CrossRef]
66. Smedby, N.; Quitzau, M.B. Municipal governance and sustainability: The role of local governments in promoting transitions. *Environ. Policy Gov.* **2016**, *26*, 323–336. [CrossRef]
67. Berglund-Snodgrass, L.; Högström, E.; Fjellfeldt, M.; Markström, U. Organizing cross-sectoral housing provision planning: Settings, problems and knowledge. *Eur. Plann. Stud.* **2020**, 1–21. [CrossRef]
68. Juhola, S.; Seppälä, A.; Klein, J. Participatory experimentation on a climate street. *Environ. Policy Gov.* **2020**, *30*, 373–384. [CrossRef]
69. Gohari, S.; Baer, D.; Nielsen, B.F.; Gilcher, E.; Situmorang, W.Z. Prevailing approaches and practices of citizen participation in smart city projects: Lessons from Trondheim, Norway. *Infrastructures* **2020**, *5*, 36. [CrossRef]
70. Noring, L. Public asset corporation: A new vehicle for urban regeneration and infrastructure finance. *Cities* **2019**, *88*, 125–135. [CrossRef]
71. Schultz Larsen, T.; Delica, K.N. Territorial Destigmatization in An Era Of Policy Schizophrenia. *Int. J. Urban Reg. Res.* **2021**, *45*, 423–441. [CrossRef]
72. Andersen, B.; Eline Ander, H.; Skrede, J. The directors of urban transformation: The case of Oslo. *Local Econ.* **2020**, *35*, 695–713. [CrossRef]
73. Richner, M.; Olesen, K. Towards business improvement districts in Denmark: Translating a neoliberal urban intervention model into the Nordic context. *Eur. Urban Reg. Stud.* **2019**, *26*, 158–170. [CrossRef]
74. Storbjörk, S.; Hjerpe, M.; Glaas, E. Using public–private interplay to climate-proof urban planning? Critical lessons from developing a new housing district in Karlstad, Sweden. *J. Environ. Plann. Manag.* **2019**, *62*, 568–585. [CrossRef]
75. Gustavsson, E.; Elander, I. Sustainability potential of a redevelopment initiative in Swedish public housing: The ambiguous role of residents' participation and place identity. *Progress Plan.* **2016**, *103*, 1–25. [CrossRef]
76. Stopper, M.; Kossik, A.; Gastermann, B. Development of a sustainability model for manufacturing SMEs based on the innovative doughnut economics framework. In Proceedings of the International MultiConference of Engineers and Computer Scientists, Hongkong, China, 16–18 March 2016; pp. 16–18.
77. Hartmann, T.; Jehling, M. From diversity to justice—Unraveling pluralistic rationalities in urban design. *Cities* **2019**, *91*, 58–63. [CrossRef]
78. Shahab, S.; Hartmann, T.; Jonkman, A. Strategies of municipal land policies: Housing development in Germany, Belgium, and Netherlands. *Eur. Plann. Stud.* **2021**, *29*, 1132–1150. [CrossRef]
79. Durán, A.P.; Green, J.M.H.; West, C.D.; Visconti, P.; Burgess, N.D.; Virah-Sawmy, M.; Balmford, A. A practical approach to measuring the biodiversity impacts of land conversion. *Methods Ecol. Evol.* **2020**, *11*, 910–921. [CrossRef]
80. Evans, M.C.; Carwardine, J.; Fensham, R.J.; Butler, D.W.; Wilson, K.A.; Possingham, H.P.; Martin, T.G. Carbon farming via assisted natural regeneration as a cost-effective mechanism for restoring biodiversity in agricultural landscapes. *Environ. Sci. Policy* **2015**, *50*, 114–129. [CrossRef]
81. Shen, L.-Y.; Ochoa, J.J.; Shah, M.N.; Zhang, X. The application of urban sustainability indicators–A comparison between various practices. *Habitat Int.* **2011**, *35*, 17–29. [CrossRef]
82. Petersson, O. *Statsbyggnad: Den Offentliga Maktens Organisation*; Studentlitteratur: Lund, Sweden, 2018.
83. Broman, G.I.; Robèrt, K.-H. A framework for strategic sustainable development. *J. Clean. Prod.* **2017**, *140*, 17–31. [CrossRef]
84. Mang, P.; Reed, B. Regenerative development and design. *Sustain. Built Environ.* **2020**, 115–141. [CrossRef]
85. Winston, N.; Eastaway, M.P. Sustainable housing in the urban context: International sustainable development indicator sets and housing. *Soc. Indic. Res.* **2008**, *87*, 211–221. [CrossRef]
86. Burton, E. Housing for an urban renaissance Implications for social equity. *Hous. Stud.* **2003**, *18*, 537–562. [CrossRef]
87. Chau, C.K.; Yik, F.; Hui, W.; Liu, H.; Yu, H. Environmental impacts of building materials and building services components for commercial buildings in Hong Kong. *J. Clean. Prod.* **2007**, *15*, 1840–1851. [CrossRef]
88. Bibri, S.E.; Krogstie, J.; Kärrholm, M. Compact city planning and development: Emerging practices and strategies for achieving the goals of sustainability. *Dev. Built Environ.* **2020**, *4*, 100021. [CrossRef]

 *sustainability*

Article

Only for Citizens? Local Political Engagement in Sweden and Inclusiveness of Terms

Bozena Guziana

School of Business, Society and Engineering, Mälardalen University, 722 20 Västerås, Sweden; bozena.guziana@mdh.se

Abstract: In both policy and research, civic engagement and citizen participation are concepts commonly used as important dimensions of social sustainability. However, as migration is a global phenomenon of huge magnitude and complexity, citizen participation is incomplete without considering the political and ethical concerns about immigrants being citizens or non-citizens, or 'the others'. Although research on citizen participation has been a frequent topic in local government studies in Sweden, the inclusiveness and exclusiveness of terms used in the context of local political engagement, which are addressed in this article, has not received attention. This article examines the Swedish case by analyzing information provided by the Swedish Association of Local Authorities and by websites of all 290 municipalities as well terms used in selected research publications on local participation. Additionally, this article studies the effectiveness of municipal websites in providing information to their residents about how they can participate in local democracy. The results show that the term *citizen* is commonly and incorrectly used both by local authorities and the Association. The article concludes that the term *citizen* is a social construction of exclusiveness and the use of the term *citizen* should be avoided in political and civic engagement except for the limited topics that require formal citizenship.

Keywords: local political engagement; citizen; citizenship; resident; inclusiveness; exclusiveness

Citation: Guziana, B. Only for Citizens? Local Political Engagement in Sweden and Inclusiveness of Terms. *Sustainability* **2021**, *13*, 7839. https://doi.org/10.3390/su13147839

Academic Editors: Ingemar Elander and Marc A. Rosen

Received: 26 May 2021
Accepted: 3 July 2021
Published: 13 July 2021

Publisher's Note: MDPI stays neutral with regard to jurisdictional claims in published maps and institutional affiliations.

Copyright: © 2021 by the author. Licensee MDPI, Basel, Switzerland. This article is an open access article distributed under the terms and conditions of the Creative Commons Attribution (CC BY) license (https:// creativecommons.org/licenses/by/ 4.0/).

1. Introduction

"Democracy Day" is a yearly event organized in Sweden by Swedish Association of Local and Regional Authorities (SALAR). While attending this event in 2018, an elected official from the municipality of Vänersborg explained that the word citizen is outdated: "We have members of the council who are not Swedish citizens. There are those who feel excluded when the term 'citizen dialogue' is used instead of, for example, 'resident dialogue'. In my opinion 'dialogue' should be enough". This conversation triggered my interest in the inclusion–exclusion dimension of participation.

In the broad sustainability discourse, concepts such as 'place identity', 'physical and social integration', and 'participation' are common buzzwords signifying different, though often overlapping, targets for policy and research [1,2]. In this context of 'sustainabilities' and regardless of exact specification, local authorities are crucial actors stating their quest for healthy, equitable, and economically sustainable communities. Many local initiatives taken under the sustainability flag have a strong flavor of deliberation, communication, dialogue, and consensus, thus implying civic engagement is a key dimension in the implementation of sustainable development [3,4]. Thus, 'citizen participation'/'public participation' is commonly considered crucial for achieving 'social sustainability' in an urban/local context [5–7]. It is also argued that participation is important for successfully monitoring social development goals [8].

This article addresses the encouragement of public political engagement at the local level, including the inclusiveness aspects of terms used in targeting the public and for labeling instruments for participation. This article also provides comprehensive information about different forms of participation. Sweden provides an interesting case because it is characterized by a high level of digitalization, a long tradition of development of citizen participation, and a high percentage of residents with foreign backgrounds.

Participation emphasizes the importance of citizens being active, not only at the time of elections but also in the intervals between elections. Researchers find that participation increases people's political self-confidence, their trust in the political system, and their understanding of the common good [9–14]. There is also a widespread agreement that including citizens will increase both the efficiency and legitimacy of government. Citizen participation is therefore loudly praised by decision-making authorities at all levels, national and local, and even the transnational level, such as the EU.

Parallel to the long-standing interest in citizen participation, the world-wide number of refugees and people in refugee-like situations is increasing exponentially. The UN Refugee Agency estimates that there were 80 million Forcibly Displaced People worldwide at mid-2020 [15]. Sweden has a long history of immigration. Recent immigration peaked in 2016 [16], which brought rapid changes in the population structure and especially to the growing number of residents who are not citizens in a legal sense. These changes have been noticed by the Swedish Contingency Agency (MSB).

During Emergency Preparedness Week in June 2018, the Swedish Contingency Agency sent out the brochure *If Crisis or War Comes—Important information for the population of Sweden* (Om krisen eller kriget kommer -Viktig information till Sveriges invånare) to all households in Sweden [17]. This brochure is now available online, with translations into three of Sweden's five minority languages (Finish, Meänkieli, and Sami) and into other languages such as Arabic, English, Farsi, French, and Russian, as well as a simplified Swedish version. The objective of this brochure was to prepare the people who live in Sweden for the consequences of anything from serious accidents such as extreme weather and IT attacks to—in the worst-case scenario—war. Similar communications were distributed in 1943, 1952, and 1961. Common to all these earlier editions was the reference to war, whereas the current edition (2018) states 'crisis or war'. More striking is the change in the terms used for the target group: people living in Sweden. In all earlier editions (1943, 1952, and 1961) *citizens* were explicitly addressed, while the edition from 2018 addresses *residents* in Sweden. This change in the term used in targeting people in Sweden is an example of using more inclusive language at the national level.

How are 'we' as people currently living in Sweden addressed in the context of political engagement and participation by authorities and by scholars at the local level? The present article argues that there is a need to discuss the use of the term 'citizen' as a crucial issue concerning who is included or excluded in a context when formal citizenship is not relevant. As many residents in Sweden's municipalities are not citizens in a legal sense, a growing percentage of constituents can be excluded by the terminology used by local governments on their websites.

Moreover, there are many different forms of citizen participation as a result of "participatory engineering [18] and the 'participatory revolution' [19]. Local authorities may increase public engagement by including an overview on their websites of the various participation tools that are available in their municipalities. As [20] (p. 25) pointed out, 'local political leaders in Sweden are the most supportive to party-based electoral democracy—and the most critical of participatory democracy—in Europe.' Therefore, it is especially interesting to investigate the comprehensiveness of information about opportunities for local political engagement provided on websites of local governments in Sweden.

Three following questions, not explicitly addressed by the body of literature, are posed in this article:

1. Are the local authorities encouraging political participation by giving comprehensive information about different tools for participation and influence?
2. Is the Association of Local Authorities and Regions taking leadership for adjusting democracy at the local level to the new reality in which an increasing number of residents in Sweden are not citizens in a formal sense?

3. Are the local authorities using inclusive terms (resident) or exclusive terms (citizen) on their websites?

By studying these questions, the article contributes to the literature in three ways. Firstly, by focusing on the vocabulary used in democracy at the local level, the paper contributes to the literature on political engagement and social inclusion, especially regarding immigrants. Political participation is regarded as crucial for integration [21,22]. The use of inclusive or exclusive terminology in local democracy can influence political integration of non-citizen immigrants. Smith and Ingram [23] draw attention to ways that social groups can be constructed by policymakers in positive or negative terms. Clyne [24] highlights an important example of this language of exclusion that was used to divide the population of Australia into 'us' and 'them.' Lane [25] found a similar phenomenon in Norway, where the use of the exclusive term 'ethnically Norwegian' as a criterion of national identity led to 'heated debate' in media about the need for more inclusive identity categories suited for a multilingual and multicultural society. Recently, Barcena, Read, and Sedano [26] published their findings that show that inclusiveness of language in Language Massive Open Online Courses (LMOOC) of elementary Spanish for refugee migrants has a positive effect not only on migrants' language learning but even on social inclusion.

Secondly, the article contributes twofold to the literature about local digital democracy. Research shows that municipal websites can empower monitoring and participating in local governments [27] and that online information can also mobilize individuals for participation offline [28]. Notably, despite the growing number of e-democracy empirical studies, scholars and local leaders have shown little interest in the comprehensiveness of information on municipal websites that promotes both online and offline participation. The article fills this gap by studying the information on Swedish municipalities' websites. The high level of digitalization [29] and a long tradition of citizen participation make Sweden suitable for such study.

Thirdly, this article offers a novel methodology of using terms as indicators for inclusiveness. Within e-government research [30] and within practitioners' work [31] inclusiveness is often addressed as accessibility in the context of digitalization. This article broadens the view of digital inclusiveness in the time of growing migration by recognizing that minorities can be excluded by the information that is made accessible to them.

The remainder of this article is structured as follows. Section 2 describes the relationship between local governments' websites and democracy. Section 3 discusses citizenship and participation in a normative theory context. Section 4 outlines the development of citizenship regulation and lists instruments of participation in democracy at the local level in Sweden. Section 5 examines the Swedish case: information and materials provided by both the main actors within local participation, such as SALAR and local governments, as well as selected research publications on participation at the local level. Section 6 concludes that the term *citizen* is a social construction of exclusiveness and should not be used in the context of any public participation or civic engagement that does not require formal citizenship.

Method and Material

This article examines the inclusiveness of terms used by municipal webpages as well as information and material provided on the webpage of SALAR. Municipal governments and SALAR are Sweden's primary actors that facilitate political engagement at the local level. Additionally, this article reviews the terms used in selected research publications regarding local participation before and after Sweden's peak migration in 2016.

Digital inclusiveness is often addressed as accessibility, both within e-government research [30] and by practitioners [31]. This narrow understanding of accessibility is a response to the public nature of local authorities' responsibilities; '[u]nlike organizations in the private sector, government agencies have a charge to make their information and services available to everyone' [32] (p. 133). A growing number of residents in Swedish municipalities are not citizens in a legal sense. Therefore, this article studies another dimension of inclusiveness: the terms that municipalities use to target individuals when facilitating information about political and civic engagement on their websites.

Municipality is the common legal label for the 290 local self-government units in Sweden, all of which currently provide information on their websites. At the time this study was conducted, the information on these websites was usually structured under 6–8 main headings, which correspond with public services operated by the municipalities such as schools, child and elderly care, utilities, housing, cultural and leisure activities, etc. Under these *main headings*, more information is available under a lot of *subheadings*, as shown in the Figure 1.

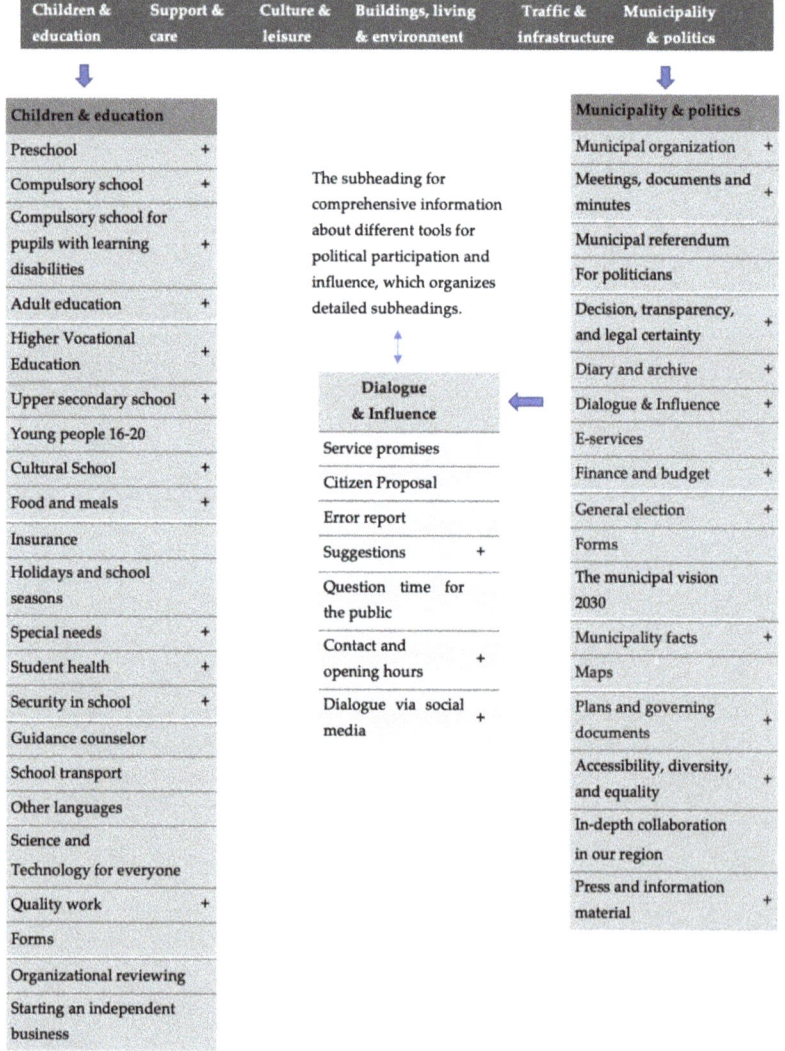

Figure 1. Example of main headings and subheadings on a municipal website.

Municipalities can make their information about means of participation easy to find by providing a comprehensive subheading, which organizes and displays the detailed subheadings about specific opportunities for residents to participate. My findings below suggest that a percentage of municipalities have already achieved this level of accessibility.

Content analysis been conducted on all municipal government websites between 1 October 2017 and 6 January 2018 using an evaluation questionnaire (see Table 1).

Table 1. The factors for analysis of the municipal websites.

Information about Participation	Inclusiveness of Language
The absence of a subheading for comprehensive information about different tools for political participation and influence at the local level under the subheading about the local government and politics (as shown in Figure 1).	The term used (citizen or resident): * under websites' subheading for means of participation and influence * when the authority offers dialogue as an opportunity for participation and influence.

It is important to keep in mind the complexity of a municipality. A municipality can be seen as a geographic entity, an organization, and a political institution [33]. In relation to their residents, a municipality acts often as a service provider and as an authority. Residents are expected to meet public authorities and participate in politics, be taxpayers, voters, employees in municipalities, and users of the public activities that have expanded as Sweden's welfare state has evolved [34]. The information provided by the municipalities on their websites must meet this complexity. For example, to meet the growing interest among local authorities in the user role of the residents [35], the websites of Swedish municipalities commonly offer a function for accepting complaints as well as other suggestions about their services. The present paper focuses on the information related to the local political participation, i.e., the political agency of the residents. The user-related information and functions available on the municipal websites are therefore excluded.

2. Local Governments Websites and Democracy

Local communities and municipalities are crucial for the development and maintenance of democracy. They play an increasingly important role in our everyday lives. Research shows that people are closely connected to the local community and tend to be more interested in their own neighborhood or municipality than to their region or the whole country. Researchers also claim that the same pattern of local connection can be seen in the use of the internet [36]. 'Although the Internet is often seen as one of the big examples of McLuhan and Power's (1989) 'global village' concept, people using these global technologies often do this on a local level' [36] (p. 6). Many of the applications and information retrieval concern the local level.

The fast development of digitalization and e-governance is creating a growing interest in local authorities' websites. Earlier examples of studies have focused on more general questions such as *public involvement* [37], *e-participation* [38] and *e-government* [39]. More recently-conducted studies focus on more specialized issues, such as *technology acceptance* [40]; *local government transparency* (Portugal) [27]; *different determinants of adaptation* (Turkey) [41], (Norway) [42]; *information quality* [43]; *e-government evaluation models* (Greece) [44,45] and *use of social media,* for example, Italy and Spain [46], South Africa [47,48], and Western European municipalities [49].

In the beginning of the digital age, many scholars and practitioners thought that the new ICTs would contribute to democracy by connecting citizens with politicians and policy makers. This potential to enable citizens to communicate directly with government remains largely unrealized. The local authorities' websites tend to provide ample public information: contact information for public officials, descriptions of the activities of municipal departments, online council agenda minutes, and downloadable forms; thus, a kind of 'billboard' of information [28]. Here, the difference between *e-government* and *e-democracy* is relevant [50]. Generally, e-government deals with the passive provision of information and online services to individuals and businesses. By contrast, e-democracy offers more active forms of public participation and engagement in decision making (for example [39,44,51]. Using this distinction, local authorities seem to be more interested in *e-government* than *e-democracy*; they primarily link the advantages of ICT with municipal service provision, for example, local authorities' websites provide information about services and self-service [52,53].

Steyaert [36] has studied information on Flemish municipalities' websites with regard to residents' different roles in the relationship to their municipality. Two of these roles involve political engagement: the role as a voter and the role as an active citizen. The third role is as a consumer or a client. The results of this study show that the municipalities tend to reduce the residents to a consumer or client of the services of the municipalities. 'The more political role of the resident, as a voter and especially as an active citizen, are not supported by the municipalities or even completely ignored' [36] (p. 15). It should be noted that Steyaert's research was published 20 years ago.

Municipal websites have changed since the beginning of the millennium. Now, Swedish municipalities, like municipalities in general, address residents in their political role with varying success. Still, population changes and new instruments for participation pose new challenges in addressing political engagement through local governments' websites. In the Swedish context, Lidén [54] has studied the supply of e-democracy on all municipalities using data between 2007 and 2009 as well as citizens' demand for e-democracy. The use of social media by municipalities has been studied by Klang and Nolin [55], Larsson [56], and Lidén and Larsson [57]. None of these previous studies has drawn attention to use of inclusive language on local government websites or to the comprehensiveness of the information these websites offer. The present article aims to fill these gaps.

3. Normative Theory Context: Citizenship and Participation

People have multiple social identities such as consumers, individual personalities, employees, members, and citizens. The role of being a citizen is confusing as citizenship is a widely contested concept (see [58] for a compilation). The literature presents different conceptualizations and dimensions of citizenship. Important examples include *civil, political,* and *social citizenship* [59]; *state* and *democratic citizenship* [60]; and, more recently, *digital citizenship* [61], *environmental citizenship* [62], and *local (urban) citizenship* [63].

Many studies explain the components of citizenship as (i) *legal status*, (ii) *political agency*, and (iii) *membership in a political community* [38,64,65].

When focusing on legal status and membership in a political community, citizenship is primarily a state-centered function; it includes national belonging with its associated rights and obligations. Britannica provides the following definition of citizenship [66]:

> [...] relationship between an individual and a state to which the individual owes allegiance and in turn is entitled to its protection. Citizenship implies the status of freedom with accompanying responsibilities. Citizens have certain rights, duties, and responsebilities that are denied or only partially extended to aliens and other noncitizens reside ng in a country. In general, full political rights, including the right to vote and to hold public office, are predicated upon citizenship. The usual responsibilities of citizenship are allegiance, taxation, and military service.

Accordingly, an individual can legally be a citizen of a state; individuals are not citizens of other administrative entities, such as municipalities, communities, or cities. Scholars identify two contradictory trends in the development of national citizenship: the ability of migrants to gain citizenship is becoming less challenging in some ways but also more challenging in other ways.

On the one hand, some political actors push for a liberalization of access to citizenship [67]. Examples include improved and simplified opportunities to obtain citizenship through registration; simplified naturalization rules, including the current period of residence; and recognition of multiple citizenship as some countries permit dual citizenship.

On the other hand, 'in the recent era of transnational population movements, national governments have harnessed both new and existing instruments to (re)assert state authority over the regulation of membership' [68] (p. 1153). Several states have considerably tightened access to citizenship and to permanent residence, and attitudes towards non-citizens have hardened [69]. For example, in the UK, the right of asylum seekers to receive the same benefits as settled citizens has been removed and replaced with reduced benefits [70].

While the path to citizenship is becoming less accessible, the value of citizenship is becoming more significant. As Joppke [67] pointed out, in the time of mass immigration, conflicts surrounding citizenship focus on the original meaning of citizenship as state membership. The tightening and loosing of citizenship rules are discussed by Mouritsen [71] in light of the literature on citizenship in sociology and political science representing 'postnational' critique and the rise of a global human rights regime [72]. The post-nationalization can be seen as a 'banalization' of the status of citizenship', and the new citizenship recognition discourse and policy can be seen as a "way of denying, resisting or reversing this post-national 'banalization'" [71] (p. 91). Furthermore, the 'banalization' of the status of citizenship is associated with two distinct developments [71]. One of them is the already mentioned liberalization of citizenship acquisition, the second one is the diminished importance of "material content and consequentiality of membership". Denizenship [73] is an example of the latter development, arguing that a state grants certain economic, social, and (sometimes) partial political rights to long-term residents who are settled within the state's borders but do not possess its citizenship. Elena Dingu-Kyrklund [74], in her paper on citizenship, migration, and social integration in Sweden, described that in the Swedish context, a resident enjoys almost the same rights as citizens in social, economic, and political terms, with some important exceptions. Thus "the citizenship issue, to a large extent, has been a secondary issue; the main and most difficult concern for non-Swedes remains that of immigration, which involves basic admission to and becoming officially domiciled in the country" [74] (p. 3).

The renewed significance of citizenship is illustrated by states attempting to differentiate more between the value of citizenship and mere residence, and in some countries between permanent and temporary residence. Hansen [75] stressed that the material and subjective value difference between citizenship and denizenship—permanent residence—in fact remains considerable and may be increasing. Ten years after Hansen's work, Hegelund [76] provided additional examples from Scandinavian countries of increasing differences between the rights of citizens and residents.

Moreover, not only *having citizenship* but *even having the right citizenship* becomes important. Research focused on global inequalities emphasizes the importance of location and its relationship with citizenship [77]. Citizenship and national location are the major factors behind differences in individual income across the globe [78,79], and the rise of a strategic-instrumental approach towards access to national citizenship [80,81] is not surprising. By strengthening the difference between citizens and non-citizens, the COVID-19 pandemic has influenced uses and meanings of citizenship in different ways, including creating problems and strategic choices for individuals who hold multiple citizenships.

In addition to the emerging importance of membership, which increases differences between citizens and non-citizen residents, nations also face the problem of the growing number of stateless people. Millions of people are stateless, and an estimated 1.1 billion lack legal identity documentation [82]. Since they are stateless, these individuals lack state acknowledgement and are 'often denied access to basic rights as education, healthcare, freedom of movement, and access to justice' [83] (p. 9). Scholars argue that territorial presence, not recognized national membership, should be the basis for migrants who are claiming rights [84,85] and should even be the basis for defining citizenship [86].

Citizenship is important for some forms of political agency. Voting and getting elected in national elections are examples of the privileges restricted to citizens. While adult citizens are entitled to vote in national elections even if they do not reside in the state, noncitizens may not vote in national elections even if they do reside in the territory of the state. The decline in voting turnout in most advanced industrial countries [87,88] has led leaders and scholars to focus on activating citizens between elections. The concept of *citizen participation* has received much attention from different fields of study and is loudly praised by decision-making authorities (see Bobbio [89] for an overview of different arrangements for participation). There are various definitions of citizen participation. Verba, Scholzman, and Brady [90] defined it as any voluntary action by citizens that is more or less directly aimed at influencing the management of collective affairs and public decision making. Arnstein [91] introduced a ladder of participation, from elemental to more in-depth participation (e.g., information, communication, consultation, deliberation, and decision making) based on levels of interaction and influence in the decision-making process. Swedish scholars have contributed to understanding the governing of participatory instruments through studies of, among others, participants' motives for participation [92], idealist and cynical perspectives on the politics of citizen dialogues [93], invited participation under pressure in a local planning conflict [94], and the rise of e-participation initiatives in non-democracies [95]. In addition, Hertting and Kugelberg [96] shed light on the problem of institutionalizing local participatory governance in relation to representative democracy.

The broad attention for strengthening participation is described by scholars as 'participatory engineering' [18] and 'participatory revolution' [19]. Besides the meaning of membership of a state, the term *citizen* is therefore commonly used for the political agency in the context of political and civic engagement; some scholars are even pointing out 'citizenship turn': '[w]hatever the problem—be it a decline in voting, increasing numbers of teenage pregnancies, or climate change—someone has canvassed the revitalization of citizenship as part of the solution' [97].

When most residents in countries were also citizens in a legal sense, the use of the terms citizen and *citizen participation* was rather unproblematic. Now, however, no part of the Earth is unaffected by migration, either as a country of origin or as a country of destination. Therefore, when scholars or government authorities use the term *citizen* to imply political beings with certain rights and obligations, this should raise serious concerns.

4. Citizenship and Local Participation in Sweden

4.1. Citizenship

For a long time, Sweden has been counted as a state with one of the more inclusive citizenship regulations [74,98]. This legislation has evolved over time in close cooperation with other Nordic countries; it started at the end of the 19th century and continued after the Second World War. For example, these joint discussions on citizenship led to adaptation of new citizenship laws in Denmark, Norway, and Sweden in 1950 [98].

Both EU-integration and globalization influenced a number of modifications that were made to Swedish legislation; among others, the positions within the Swedish bureaucracy and judiciary that should be restricted to citizens were reduced to very few positions. Only some top positions in the judicial system (judge, prosecutor) and some few top administrative positions still require Swedish citizenship. Others, such as ordinary judicial and related positions (lawyers, jurors) that previously had required Swedish citizenship, are now no longer restricted. This inclusiveness towards non-citizen residents separated Sweden from other Nordic countries; the revised Citizenship Act, adopted in 2001, was the first codification that was not based on Nordic cooperation [74]. The liberalization of citizenship rules has been more far-reaching in Sweden than in other Scandinavian countries [98]. For example, acceptance of dual citizenship was introduced in Sweden almost two decades before Norway [99]. The main landmarks in the development of Swedish citizenship legislation are presented in Table 2. The Swedish legislative process begins with the appointment of a specialized committee that is given a mandate to investigate the question to be legislated. This committee produces a report with a suggestion for new legislation, a so-called SOU or Ds. SOU—Sveriges Offentliga Utredningar (Sweden's Public Investigations); Ds—Departements serien (Ministry series, a smaller, shorter form of public investigations handled within the ministry in question) [74].

Table 2. Development of legislation of Swedish citizenship with a focus on acquisition of citizenship and participation. Partly based on [74,98].

Year	Legislation and/or Committee Report	Effect on Acquisition of Citizenship and Participation in Sweden
1894	Citizenship Act	
1924	Citizenship Act	New naturalization conditions: (1) be over 21 years of age, (2) have accumulated 5 years of domicile in Sweden, (3) have exhibited good conduct, and (4) demonstrated a capability to provide for himself and his family.
1950	The 1950 Act (1950: 382) on Swedish citizenship	Established common immigration policies across Scandinavia that allowed selected rights to immigrants.
1976	SOU 1975:15 [100] Municipal suffrage for immigrants	The electoral reform gave electoral rights to foreign citizens who had lived at least 3 years in Sweden. The three-year waiting period for citizens from EU countries, Iceland, and Norway was abolished [101].
1976		A major reform of the naturalization rules in Sweden: the waiting time for Naturalization was shortened from 7 to 5 years, and to 2 years for Nordic citizens. The residence request for acquisition by notification was shortened from 10 to 5 years.
2001	Revised Swedish Citizenship Act (2001: 82) SOU 1999:34 [102] Swedish Citizenship	Acceptance of dual citizenship
2013	SOU 2013: 29 [103] Swedish citizenship	The definition of the meaning of Swedish citizenship: 'Swedish citizenship is the most important legal relationship between the citizen and the state. Citizenship involves freedoms, rights and obligations. It is a basis for Swedish democracy and represents a significant link with Sweden.' Annual Citizenship ceremonies for new Swedish citizens should be held by municipalities.
2021	SOU 2021:2 [104] Requirements or knowledge of Swedish and social studies for Swedish citizenship	Pending

Swedish regulations regarding citizenship ensure that a resident enjoys almost the same rights as a citizen—in social, economic, and political terms [74]. Two notable exceptions are: '[...] the right to vote in national elections and, especially the unrestrained, inalienable right to reside, which are still exclusively reserved for citizens' (p. 8).

Residents with a foreign background represent a growing share of the population in Sweden. Similarly, the number of applications for citizenship is also growing, as shown in Table 3. Foreign background includes foreign-born and domestic-born with two foreign-born parents. Before 2001, even domestic-born with one foreign-born parents was considered as foreign background [105].

Table 3. Granted application on citizenship [106].

Year	Granted Applications	Share of Population with a Foreign Background (%)
2000	data not available	14.5
2010	28,100	19.1
2011	33,112	19.6
2012	46,377	20.1
2013	46,849	20.7
2014	38,890	23.5
2015	44,209	22.2
2016	56,037	23.2
2017	65,611	24.1
2018	61,309 (* 93,261)	24.9
2019	74,924 (* 109,580)	25.5
2020	81,377 (* 119,728)	data not available

* The number of submitted applications.

Unlike many others European countries, Sweden at present has no language or civics tests for people applying for citizenship. There is, however, a good conduct clause requirement, and either a criminal record or unpaid debts can affect applications. Furthermore, applicants need to have lived in Sweden for five years, or three if they are a cohabiting partner of a Swedish citizen, before they can apply for citizenship. However, the Swedish government has launched an inquiry that investigates how the law could be changed to make it compulsory for applicants to pass a test on the Swedish language and civics to get citizenship. The final report is to be presented by 1 July 2021, with the parts of the report dealing with the language and civics tests presented 13 January 2021 [103]. The Inquiry states that the purpose of the requirement is to enhance the status of citizenship and promote an inclusive society. The Inquiry proposes that the Swedish Citizenship Act (2001:82) should stipulate that knowledge of Swedish and civics is required for the acquisition of Swedish citizenship. The knowledge requirement for Swedish citizenship should cover people who have turned 16 but not 67 years. However, this requirement should not apply to state-less persons born in Sweden who are under 21 or to Nordic citizens who acquire citizenship through the provisions on notification in Section 18 and Section 19 of the Citizenship Act. The citizenship test in civics should be based on a book specially produced for the purpose. The book should contain knowledge needed to live and function in Swedish society focusing on democracy and the democratic process and it should be available for download in Swedish and ten immigrant languages in Sweden. Additionally, the government is also looking into introducing a similar requirement for obtaining permanent residence. In sum, Swedish citizenship laws include both contradictory trends: facilitating access to citizenship while also restricting citizenship.

4.2. An Overview of Forms for Participation

The Swedish constitution states that public power in Sweden is derived from the people and there are three levels of domestic government: national, regional, and local. Sweden is divided into 290 municipalities and 21 regions. In January 2020, the county councils (landsting) of Sweden were officially reclassified as Regions (regioner). In addition, there is the European level, which has acquired increasing importance following Sweden's entry into the EU.

The 1992 Swedish Local Government Act (LGA) regulated the division into municipalities and county councils as well as the organization and powers of these municipalities and county councils. The LGA states that all residents are *members* (a person who is registered as a resident of a municipality, owns real property there, or is assessed for local income tax there is a member of that municipality [107]) in a municipality, not *citizens*. With the new 2017 LGA, the county councils were changed into regional councils. The Swedish Association of Local Authorities and Regions changed the name in Swedish (from Sveriges Kommuner och Landsting (SKL) to Sveriges Kommuner och Regioner (SKR); n English the name is SALAR, in both cases), and the county councils (landsting) of Sweden were officially reclassified as Regions (regioner) in 2019. The LGA also contains rules for elected representatives, municipal councils, executive boards, and committees. There is no hierarchical relationship between the local and the regional level since municipalities and regions have their own self-governing authorities with responsibility for different activities.

In Sweden, like most countries, *voting* is a political act, limited at the national level to members of the state. However, as it is shown in Table 2, since 1976 foreign citizens who had lived at least 3 years in Sweden had electoral rights in municipal elections and in elections to the county council assembly. This three-year waiting period was abolished for citizens from EU countries, Iceland, and Norway in 1998 [104]. Bevelander and Spång [108] refer to EUDO-Citizenship Observatory 2016 and emphasize that Sweden, Denmark, and Finland are the most inclusive countries in Europe when it comes to voting rights for non-EU citizens). Since 1970, elections for local and regional level representatives have been held on the same day as the general election in Sweden.

The opportunity for public participation in planning and building processes has a long-standing tradition and is obligatory and meticulously regulated in Sweden and other Nordic countries [109]. However, this paper has been exclusively focused on other forms of participation. It can be added that in the Swedish Planning and Building Act the term 'citizen' is not used, only 'resident' (boende).

In the beginning of this millennium, researchers and policy makers promoted new initiatives to activate citizens between elections in Sweden [110] partly due to recommendations from the first Commission on democracy. The final report of the Commission, 'Sustainable Democracy. Policy for the Government by the People in the 2000s' included a number of proposals concerning local democracy and also stated that 'every citizen must be afforded greater opportunities for participation, influence and involvement.' [111] (p. 243). Furthermore, the Government Democracy Bill from 2001 [112] declared democracy as a policy area of its own and encouraged the 'municipalization of democracy' [113] (p. 136). The need for democratic expertise emerged along with the view that democracy is an issue not only for political parties but also for the municipalities. A growing number of municipal officials, 'democracy operators' [113] (p. 87) work to promote local democracy with support from SALAR.

Table 4 depicts an overview of the formal instruments of participation in democracy at the local level in Sweden. *Voting, contacting politicians* and *attending public questions time at council meetings* are examples of traditional political acts at the local level. *Citizen dialogue* and *e-petition* are examples of tools for participation. In 2006, the Swedish Association of Local Authorities and Regions initiated a project to support both citizen dialogue and e-petition as vital tools for civic engagement [114]. This Association has actively promoted the citizen dialogue by producing a great deal of published information, working with networks, conferences and awards, etc., thus acting as a 'policy entrepreneur' [93]. In 2015, 83 percent of Sweden's municipalities and county councils (stated that they had implemented some form of citizen dialogue [115].

Table 4. Examples of formal instruments of participation in democracy at the local level in Sweden.

Instrument	Requirements
National election	Swedish citizen aged 18 or more, who is or has been registered in Sweden
Voting (local election)	Citizens of other EU countries, Iceland, or Norway who are registered in the municipality or county Citizens of other countries who have been registered in Sweden for a minimum of three years and are registered in the municipality
Contacting politicians or local government officials	none
Public question time at a council meeting	none
Participation in citizen dialogue/resident dialogue (medborgardialog/invånardialog)	none
Proposing or signing a people's initiative (folkinitiativ) (instrument for direct democracy) Introduced 1994, strengthened 2011	Residents with voting rights in the local election
Submitting a "citizen proposal" ((medborgarförslag) * Introduced 2002 This instrument has been regulated in LGA, which gives it a special position among other participatory instruments)	Members of the municipality, i.e., registered in the municipality
Proposing or signing an e-petition *	Varied; residents (Västerås), registered members (Haninge), everyone (Borås)

* depending on the availability of the instrument.

A common goal among the different forms of participation is activating voters between elections. Another goal is to involve in local politics those without a legal right to vote, such as children or residents with foreign backgrounds. In 2002, an instrument called *citizen proposal* was introduced to target both goals. For local authorities that decided to implement this instrument, the *citizen proposal* (CP) enables all residents who are registered in a municipality, including those without a legal right to vote, to raise issues to the local government regarding local areas of responsibility. The CP process was included in the Local Government Act (LGA) on 1 July 2002, which gives this tool a special position among participatory instruments. Other instruments, such as e-petitions and citizen dialogues, produce suggestions that local representatives are not obliged to take into account. According to the LGA, the *citizen proposal* should be processed to enable the council to make a decision within one year of the date on which the citizen proposal was tabled, and the processing of the CP should be described in the standing order for the assembly. This instrument has spread to more than half of Swedish municipalities. In 2016, 188 (65%) of Sweden's 290 municipalities had information on their websites about how citizens can submit a citizen proposal [116].

Instruments for *direct democracy* are relatively weak in Sweden. Although Sweden has been ranked as the 'most democratic country' in the world [117], there are limited formal rights for direct civic participation [118]) and Sweden is instead characterized by a lack of 'referendum culture' [119]. For example, since introduction in 1921 only 6 national referendums have been held. Since 1977, institutionally initiated referendums have been allowed to be held at local and regional levels [119], which have been regulated in the LGA. The regulation of local and regional referendums has been changed twice during this time. Due to inspiration from Finland, the *people's initiative* has been introduced with the constitutional amendment of 1994 and in the new Municipal Referenda Act [120]]. The *people's* initiative gives residents with voting rights the ability to initiate a referendum process by getting 5% of the population in a municipality to sign a petition. However, the referendum would only be enacted if a majority of the municipal assembly approved of the referendum. In practice, very few initiatives were forwarded by the local authorities to the electorate for a consultative popular vote. The dysfunctionality of this tool led Sweden to strengthen the *people's initiative* with the constitutional amendment of 2011: only if 2/3 of a municipal assembly opposed any specific people's initiative could the referendum be denied. This has led to several policy changes initiated by people referendums [121]. By 2018, 174 referendums initiated by people in 105 of 290 municipalities have been enacted [122], i.e., the majority of municipalities have not yet achieved this form of participation.

In various international rankings, Sweden appears among the most advanced OECD countries in terms of the level of digitalization of its society and economy. In the Digital Economy and Society Index 2018, Sweden 'ranks second regarding the use of Internet by its citizens, third in terms of the use of Internet for transactional services (including banking and shopping), and third in terms of individuals' use of the Internet to send filled forms to public authorities' [29]. As there are different forms of participation as well as other online options for giving suggestions to the municipality, local leaders can simplify participation for their residents by providing an overview of participation tools using inclusive language. Local government websites that provide clear, inclusive information about participation can help residents to get involved in local politics. These aspects of the online information of the municipal websites are studied in the present paper.

5. Terminology in the Swedish Case

5.1. Previous Research

As participation at the local level has gained attention from politicians and practitioners, different research projects in Sweden have studied participation. This section provides a review of the terms used in examples of research publications before and after 2015 (see Table 5). The year 2015 'was characterised by very strongly increasing numbers of people seeking protection. In total, Sweden registered almost 163,000 new asylum applicants, more than twice as many as during the year before, which had already marked a record' [123] (p. 4). In all the studied publications, both the term 'resident' and 'citizen' are used.

Two of these publications have 'residents' or 'residents dialogue' in the title. One of them, 'The future is already here. How residents can become co-creators in the city's development' [124] is a book published by a research project about the interplay between citizen initiatives (medborgarinitiativ) and invited participation in urban planning. Despite including 'residents' in the title, most of chapters of this publication use the term *citizen* as well as *citizenship* for the political agency. However, in two chapters written by Stenberg [124] the term *resident* is used, except in the general expression 'citizens role in the planning' in the title of one of these chapters, and the introduction to the second chapter.

The other publication, published within a project about justice and socially sustainable cities, is titled 'The role and forms of resident dialogue The Västra Götaland region's consultation with civil society' [125]. In this publication, the term 'resident dialogue' as well as 'resident' are used considerably more frequently than 'citizen dialogue' and 'citizen', but still not consistently and without any reflection on differences between these terms. In a previous publication in the same project [126], the use of the term *citizen dialogue* (30) in comparison to the term *resident dialogue* (3) was dominant. This indicates a purposeful change in the choice of the vocabulary in the publication from 2015. The lack of discussion on the differences between terms used in this publication is therefore even more striking.

In spite of those examples indicating some awareness of the terms used, an explicit discussion of the inclusiveness/exclusiveness of the terms is lacking. Only two short footnotes are included that address terminology: one footnote states that *citizens* and *citizen dialogue* 'should not be understood in a narrow legal sense but as synonymous with those who live, stay and work in the city' [127] (p. 4); and the second footnote highlights 'the linguistic challenge of using the term *citizen* in the time when more and more people are living as non-citizens' [128] (p. 77).

Notably, there are no considerations on inclusiveness of language in the newly published special issue of the Swedish planning journal Plan [128] on 'Planning and democracy'. The contributions by established scholars on participation and local governance do not draw attention to resident terminology. 'In PLAN, researchers and practitioners describe, analyze and debate changing conditions and new challenges for community planning, new working methods and the development of the profession. PLAN monitors municipal and regional planning, regional policy, construction and housing policy, infrastructure, environmental policy and international development trends, and highlights social consequences in planning. PLAN makes room for new theoretical perspectives and young writers!' [129].

By contrast, Wiberg [130] in her article 'The political organization of citizen dialogue' published at Stockholm center for organizational research (SCORE) mentioned the exclusiveness of the term *citizen dialogue* and inclusiveness of the term *resident dialogue* as not all residents are Swedish citizens. Wiberg also pointed out that several municipalities (p. 7) have started to use the term *resident dialogue* instead of *citizen dialogue*, e.g., Halmstad municipality.

Table 5. Terms used in selected examples of publications on local public participation.

	Stenberg et al., 2013 [124]	Olofsson, 2015 [127]	Abrahamsson et al., 2015 [125]	Jahnke et al., 2018 [131]	Plan 2021 [128]
Title	'The future is already here. How *residents* can become co-creators in the city's development'	'A research-based essay on the possibilities and obstacles of dialogue'	'The role and forms of *resident dialogue* The Västra Götaland region's consultation with civil society'	'Management system for better *citizen dialogue*. A study of the management and organization of dialogue work in the public construction sector'	'Planning and democracy'
Resident *	60	27	84	10	14
Citizen *	253	26	17	1	40
Resident dialogue	1	1 (references)	26	-	-
Citizen dialogue	73	36	7	21	117
Reflection about terms used	–	Footnote in Summary: Citizens and citizen dialogue should not be understood here in a narrow legal sense but as synonymous with 'those who live, stay and work in the city' (p. 4)	–	–	Footnote in chapter by Ingemar Elander: Even the word 'citizen' is a linguistic challenge today when more and more people are living as non-citizens within the national border, they happen to be in. 'Refugees', 'asylum seekers', 'undocumented', 'unaccompanied minors', 'illegal', 'irregular', 'the others', 'nomads', and 'invisible' are some of the different names. (p. 77)

Table 5. Cont.

	Stenberg et al., 2013 [124]	Olofsson, 2015 [127]	Abrahamsson et al., 2015 [125]	Jahnke et al., 2018 [131]	Plan 2021 [128]
Examples of the use of the term *resident*	* resident initiative (p. 95)	* make residents co-creators of societal change (p. 18) * resource-poor residents (p. 28)	* city residents (p. 21) * dialogue with residents (p. 30)	* resident involvement (p. 6)	* Citizen participation is not just about giving residents the opportunity to participate in digital platforms (p. 84) * dialogue between residents and planners (p. 104)
Examples of the use of the term *citizen*	* citizen initiative (p. 81) * democratic citizen participation (p. 100) * citizen partici-pation (p. 108)	* a politically active citizen between the elections (p. 11) * responsible citizens (p. 40) * the empowered citizen (p. 40)	* citizen democratic influence (p. 31) * low-abiding citizens (p. 9)	* citizen participation (p. 4)	* in our role as a citizen (p. 105) * dialogues between municipality, citizens and builders (p. 25) * concerned citizens (p. 23)
Other terms	* people (p. 23)	* People's political participation (p. 112)	* persons participation (p. 19)	* different inndividuals (p. 6)	* people (p. 8) * people's (p. 3)

5.2. The Swedish Association of Local Authorities and Regions (SALAR)

The mission of SALAR is "to provide municipalities, county councils and regions with better conditions for local and regional self-government. The vision is to develop the welfare system and its services. It's a matter of democracy." The Association is involved in promoting local democracy in Sweden, often acting as a 'policy entrepreneur' [92].

Democracy, leadership, governance (Ddemokrati, ledning, styrning) is one of the eight main headings on SALAR's website. As shown in Figure 2 SALAR uses the term *citizen*, even when targeting people who wanted more information about the LGA, initiatives, and referendums. In its publication on the strengthened 'people's initiative', SALAR refers to 'citizens entitled to vote in the municipality or county council') 'röstberättigade medborgarna i kommunen eller landstinget.') [132] (p. 2). The Association uses this *citizen* term despite the aforementioned fact that citizenship is not required in order to vote in local elections. Similarly, in a platform for discussion and standard setting "Ten factors for 'good local democracy'", SALAR explains that citizen participation is one of the factors, thus referring to municipal residents as *citizens*.

> **Medborgare i Sverige som har frågor om kommunallagen, folkinitiativ, folkomröstning**
> Vänligen tag kontakt med **din kommun eller ditt landsting eller region**.
>
> Sök kontakt- och adressuppgifter till din kommun
>
> Sök kontakt- och adressuppgifter till ditt landsting eller region

Figure 2. Targeting people who need or want more information about the LGA and instruments for local participation (own translation: Citizens in Sweden who have questions about the Local Government Act, people's initiative, referendum).

In their efforts to support municipalities and regions, SALAR conducts different surveys and comparisons. Some surveys are related to democracy issues such as the Democracy Barometer (Demokratibarometer) and Information for All (Information för alla). The survey Information for All was conducted yearly between 2009 and 2017, with around 250 questions regarding different municipal services (preschool, elementary school, high school, elderly care, individual and family care, disability care, building and living, streets, roads and environment, permits, business and more, non-profit sector, culture and leisure, and business), as well as transparency and influence, and municipal websites' search functions. SALAR reports that these surveys have 'significantly contributed to developing websites of the municipalities based on citizen perspectives' and 'the results have improved year by year' [31] [p. 5). The majority of questions asked in Transparency and Influence (see Appendix A) are about what information the website users can access; there are few questions about how the information is available or for whom (audio, sign language, other languages, web TV, etc.). There are no questions about websites' description of the means for political engagement and participation. SALAR also conducts studies on various issues, including participation and democracy, among others. Similarly, SALAR's study about the people's initiative [121,133] focused on whether websites' users could find information about specific people's initiatives that are in progress in the municipality; this study did not ask if websites explain that residents have a right to initiate a new referendum.

5.3. Local Government Websites

The entire territory of Sweden is divided into 290 local self-government units. All of these local units provide information on their websites. The results of content analysis of these websites are presented below.

5.3.1. Main Subheading for Participation and Influence

A vast majority of municipalities' websites include some heading related to local politics, often *Municipality and Politics* (*Kommun and politik*). Surprisingly, only 71 percent of municipalities provide a comprehensive subheading for the means of participation under their heading related to local politics. Table 6 depicts the topics in these subheadings that encourage residents to engage in local politics. 'Influence' and 'dialogue' are most frequently used. The choice of the term 'influence' can be explained by its explicit correspondence to agency. A possible explanation for the term 'dialogue' can be the strong establishment of 'citizen dialogue' as an instrument for participation. Notably, some municipalities choose to combine participation and making complaints about their services.

Table 6. Topics in subheadings for comprehensive information for participation and influence on municipal websites.

Topic		Share
Influence	(Påverka)	43%
Dialogue	(Dialog)	36%
Democracy	(Demokrati)	8%
Insight Access	(Insyn)	5%
Appealing a decision	(Överklaga)	1%
Others		6%

Under the subheading for comprehensive information, the local governments list different forms of participation. Which forms are listed varies. The variation partly depends on the decision by the local authority to introduce instruments such as citizen proposals or e-petition. Very few, only 5%, of all municipalities provide information about their residents' right to use the 'people's initiative'.

In some few cases, municipal websites organize the online content under this comprehensive subheading by roles instead of actions; for example, there are subheadings for councilors (fullmäktigeledamot), elected officials (förtroendevald), citizens (kommunmedborgare), and municipality residents (Hudiksval).

5.3.2. Targeting Residents

Municipalities can activate their residents by explicitly including residents and participation options on their municipal websites. As such, the analysis presented below includes two aspects of targeting residents on local websites. The first aspect focuses on how residents are targeted under the subheading for comprehensive information about participations tools. The second aspect focuses on what term is used when the authority offers dialogue as an opportunity for participation.

Under Subheading for Participation and Influence

As shown in Table 7 the term 'citizen' is used by 32 percent of municipalities. Some communities are addressing their residents as 'You a citizen' for example 'You a citizen in the city of Gothenburg'. Other communities use the term 'municipal citizen' (kommunmedborgare) (e.g., Håbo, Bengtsfors).

Table 7. Terms used to address residents.

Terms		Share
Citizen	(medborgare, kommunmedborgare)	26%
You	(du)	22%
Resident	(invånare, kommuninvånare)	21%
Mixed (both resident and citizen)		6%
None		24%

Citizen or Resident Dialogue

The term *citizen dialogue* (medborgardialog) dominates under the subheading for information about participation and influence or the main heading for local politics; only 8 municipalities use the subheading *resident dialogue* (invånardialog). Moreover, some municipalities also published a steering (policy) document on the process of citizen/resident dialogue (Bollebygd, Forshaga). Notably, in some few cases, such a document is offered even when a clear subheading for the dialogue as an instrument for participation is missing (Västerås stad, Lilla Edet, Vänersborg). The publishing of these policy documents can be the result of the SALAR's efforts to support municipalities with development of citizen dialogue as instrument for participation, where the local authorities focus on development of steering document for the use of municipal officials but fail to provide clear information for the residents.

6. Discussion: Local Political and Civic Engagement for Citizens or for Residents?

Citizenship is a political concept familiar to the majority of people that connotates membership in a particular country. The term *citizen* is also well established in the context of participation as *citizen participation*. Norris [134] has developed the concept of 'critical citizens' to describe the long-term trend of people becoming more critical to the political systems. In times when immigration and multiculturalism pose new questions about citizenship, it is time to be critical about the use of the term 'citizen' in political and civic engagement. National elections continue to require a legal distinction between citizens and non-citizen residents, but many other forms of participation do not require this divisive distinction. Citizenship is a term that is too narrow and exclusive: populations are changing, and hardening boundaries between citizen and non-citizens are contributing to growing exclusiveness in the terms *citizenship* and *citizen*. Kuwait provides an extreme example of the importance of replacing the term "citizen" with more inclusive language, as citizens only comprise 32% of the population. Mirchandani, Hayes, Kathawala, and Chawla [135] use the term "resident" instead of "citizen" in their article on preferences for e-government services and portal factors. As Bosniak [84] and Ochoa Espejo [85] pointed out, territorial presence should be more central than national membership, i.e., people should be seen as residents, not as citizens.

The use of the term *citizen participation* and *citizen* for political agency, as established both among scholars and local authorities, is no longer appropriate. Civic participation problems due to citizenship are stressed by the United Nations Department of Economic and Social Affairs [136] (p. 47) in their report about social inclusion:

> In diversified local environment, it is easier for many residents to identify with the city where they live, work and interact, rather than the national state. In policy discourse related to social inclusion, citizenship is frequently invoked . . .
>
> Citizenship, by definition, is membership of a political community and includes rights to political participation. There is a need for finding a way to allow inclusion of all residents in a particular location so that none of them are excluded or marginalized . . . A new construct of "membership" in cities may be considered as a solution.

These problems concern formal issues related to citizenship, but as the citation in the beginning of this article shows, and as pointed out by Clyne [24], words matter; language is either more inclusive or less inclusive. Some local authorities in Sweden have taken positive steps to switch their terminology from 'citizen' to 'resident' as in the case of 'resident dialogue', which is also observed by other scholars [130]. The Swedish Contingency Agency (MSB) has provided an example of inclusive language on the national level [17]. Consequently, SALAR's use of 'citizen dialogue (including publication in English (SALAR, n.d.)) and addressing residents in Sweden as 'citizens' in the context of local democracy is therefore even more astonishing.

Moreover, it is problematic that some instruments for increased political participation at the local level are labelled with 'citizenship'-related terminology, such as 'citizen dialogue' or 'citizen proposals'. Sweden introduced citizen proposals to involve those without a legal right to vote, such as children or residents with foreign backgrounds. Municipalities are using the term 'citizen proposal' and then explaining in the information about the instrument who can submit such proposals; often as 'you who are registered in' (e.g., Haparanda). However, some municipalities also tend to use the term 'citizen' in the description of the instrument, as shown in the following examples: 'Citizens, i.e., those who are listed in the municipality, can submit ('medborgare, dvs de som är folkbokförda i kommunen, kan lämna förslag på beslut till kommunfullmäktige')' (Hedemora); ' . . . via citizen proposal, you as a citizen in the municipality of Hudiksvall can have influence on that what happens in our municipality ('genom medborgarförslag kan du som medborgare i Hudiksvallkommun vara med och påverka det som händer i vår kommun')'. It can be added that the term for initiatives in Swedish is 'people'-based (folkinitiativ). In English versions of public content, both citizens' initiative and people's initiative are used. Even the Swedish term for the referendums is people-based (folkomröstning))

Citizenship can be seen as a socially constructed practice [70], and there is a growing interest in understanding citizenship's power as practice and status [137]. For example, Dominelli and Moosa-Mitha [70] examined how social workers in practice addressed issues of citizenship. In the field of participation in local politics, formal citizenship and national voting qualification are not relevant. Municipal officials acting as 'participatory engineers' [18] and 'democracy operators' [113] should therefore review their practice of citizenship as it can influence the integration of immigrants. Martiniello [21] identifies four dimensions of integrating immigrants in politics: acquiring rights; subjective identification with the host society; adopting democratic values; and, finally, political participation. Goodman and Wright [22] identified three stages of immigrant socio-economic and political integration. The first order achievement is language and/or knowledge acquisition. The second order achievement is functional navigation/meeting of immediate needs, for example navigating the health care system. The third order achievement is membership in the polity, with civic navigation/identification with the polity. The participation of immigrants in the political process is a vital aspect of the integration of immigrants into their new society, enabling immigrants to become politically represented and to obtain political equality [138].

7. Conclusions

The empirical focus of this paper was whether political engagement at the local level is supported by using inclusive language and by clarifying online information about different means for participation.

The first research question was whether local governments provide clear information on means of participation between elections. Although the vast majority of their websites have a heading about politics, few of them provide comprehensive information about different means of participation, and the information about the instrument of 'direct democracy people's initiative' is exceptionally limited. Although SALAR [132] stresses that clear information about how to participate between elections is important for getting people involved in democracy at the local level, SALAR should also encourage municipalities to provide such information on their websites.

The second research question asked if SALAR is taking leadership for the adjustment of democracy to population changes in terms of who is and who is not a citizen. Although SALAR is generally a 'policy entrepreneur' in the field of local political participation [93], they still use the term *citizen*, both as an instrument for dialogue, and when targeting the public and the residents in the municipalities. In this way, SALAR undermines its own mission. Thus, instead of helping municipalities engage all of their residents, they contribute to marginalizing non-citizens who could otherwise actively contribute to local democracy.

The third and final research question concerned inclusiveness of terms used by local authorities on their websites. According to findings, local governments address the members of their municipalities in the context of political participation in different ways, for example as 'resident' or 'you'. Still, the term *citizen* is used by one third of municipalities. Furthermore, as an instrument for dialogue the term citizen dominates; there are examples of the use of 'resident dialogue', but they are few. In short, municipal leaders are also excluding residents who have the legal right to participate in municipal actions.

Due to migration, population structures are rapidly changing in many countries and a growing number of residents are non-citizens. As a consequence, the use of the term 'citizen' in the context of any public participation or civic engagement that does not require formal citizenship can be regarded as a social construction of exclusiveness. Local government authorities and agencies, as well other actors working with local political engagement, should thus be very careful about the term used both when explaining forms of participation and when addressing their public or individuals.

Based on a study of 700 citizen dialogue projects in Swedish municipalities, a summary publication [128] paints a nuanced picture of the pros and cons of resident involvement in local politics (outside the local government formal decision-making agenda). One general conclusion is that the dialogue is a baseline for residents to have a potential influence in specific matters of policy-making. This kind of dialogue arrangement represents a kind of 'mini-publics' [139]. In line with this conclusion, SALAR should review its own use of the term *citizen*, both when addressing Swedish residents—citizens or non-citizens in a formal sense—at the local level and in its work with participatory arrangements. This association of local and regional governments should also encourage its member municipalities to use inclusive terms on their local websites and portals, especially when informing residents about opportunities for local political engagement.

The study of civic-engagement-related information in this article is supply-oriented. Future research should rather focus on the demand side, i.e., the residents. Invited participation opportunities will not have any impact unless residents are aware of their existence. Future research should explore mechanisms of how residents acquire information about participatory means available in their municipalities. For example, future research should investigate whether online municipal information is a way to increase residents' knowledge and engagement in local politics.

This article started with a citation that illustrates that non-citizen residents may feel excluded when citizen-based terms for participatory means are used. Future research should study if this feeling of exclusion is a common experience and explore whether non-exclusionary language makes any difference in terms of local people's interest in and influence on local politics. Sweden has 290 municipalities, and as demonstrated in the PLAN-study referred to above, there is great variation between dialogue projects, which issues are at stake, and how local decision-makers and invited citizens and other residents respond.

Participation could be individual or collective, legal or illegal, aiming at consensus or challenging law and order [140]. Thus, studying "mini-publics" in action must be embedded in a wider policy context and have a longer time horizon than one particular project [139] (p. 246). This complexity requires conceptually informed, in-depth studies of civic participation in specific projects and municipalities, including direct observation and interviews with 'democracy operators' [113]. Thereby, we can increase our understanding of how residents interact with municipalities such processes and give input to inspire future development of resident participation, dialogue, and influence in local politics.

Funding: This research received no external funding.

Institutional Review Board Statement: Not applicable.

Informed Consent Statement: Not applicable.

Data Availability Statement: Not applicable.

Acknowledgments: The author wish to thank the three anonymous reviewers for their valuable comments.

Conflicts of Interest: The author declares no conflict of interest.

Appendix A

Table A1. Indicators for Transparency and Influence (Öppenhet och Påverkan) in the SALAR survey on municipal websites in Sweden, conducted yearly between 2009–2017.

Issue	Accessibility "to"	Accessibility "How"
The complete budget	x	
A simplified version of budget adjusted for the citizens and target groups	x	
General information about how complaints and opinions are handled	x	
Handling of complaints and opinions	x	
Information on distribution of seats from the last election	x	
Information about coalition, alliance, and technical cooperation in elections	x	
Contact information for chairpersons of the municipal council, municipal executive board, and committees	x	
Information about the telephone number of all the politicians in the municipal council and on the committees	x	
Frequently asked questions (FAQs) are collected	x	
A search function and an A-Z index with municipalities' responsibility and contact information	x	
The complete annual report	x	
A simplified version of the annual report for the citizens of the municipality	x	
Possibility of subscribing to an electronic newsletter	x	
Information (or details of agenda, time, and place) about municipal council meetings	x	
Information (or details of agenda, time, and place) about municipal executive board meetings	x	
Information (or details of agenda, time, and place) about municipal committee meetings	x	

Table A1. *Cont.*

Issue	Accessibility "to"	Accessibility "How"
Documents for municipal council meetings before meetings have occurred	x	
Documents for municipal executive board meetings before meetings have occurred	x	
Documents for committee meetings before meetings have occurred	x	
Protocols of municipal council meetings	x	
Protocols of municipal executive board meetings	x	
Protocols of committee meetings	x	
Possibility for citizens to search in the municipality's records	x	
The website has been adapted so that it is easy to read		x
The website has information in sign language		x
Information about municipality activities can be found in languages other than Swedish (English)		x
Municipal council meetings are distributed through Web TV		x
Information about municipalities' insurance	x	
Use of social media on the Web (e.g., Facebook)		x

References

1. Gustavsson, E.; Elander, I. Behaving clean without having to think green? Local eco-technological and dialogue-based, low-carbon projects in Sweden. *J. Urban Technol.* **2017**, *24*, 93–116. [CrossRef]
2. Vranken, J.; De Decker, P.; Van Nieuwenhuyze, I. *Social Inclusion, Urban Governance and Sustainability. Towards a Conceptual Framework for UGIS Research Project*; Garant: Antwerpen Apeldoorn, Belgien, 2013.
3. Lafferty, W.M. *Sustainable Communities in Europe*; Earthscan: London, UK, 2014.
4. Roseland, M. *Toward Sustainable Communities: Solutions for Citizens and Their Governments*; New Society Publishers: Gabriola, BC, Canada, 2012.
5. Murphy, K. The social pillar of sustainable development: A literature review and framework for policy analysis. *Sustain. Sci. Pract. Policy* **2012**, *8*, 15–29. [CrossRef]
6. Eizenberg, E.; Jabareen, Y. Social sustainability: A new conceptual framework. *Sustainability* **2017**, *9*, 68. [CrossRef]
7. Mensah, J.; Casadevall, S.R. Sustainable development: Meaning, history, principles, pillars, and implications for human action: Literature review. *Cogent Soc. Sci.* **2019**, *5*, 1653531. [CrossRef]
8. Saner, R.; Yiu, L.; Nguyen, M. Monitoring the SDGs: Digital and social technologies to ensure citizen participation, inclusiveness and transparency. *Dev. Policy Rev.* **2020**, *38*, 483–500. [CrossRef]
9. Karlsson, M. Participatory initiatives and political representation: The case of local councillors in Sweden. *Local Gov. Stud.* **2012**, *38*, 795–815. [CrossRef]
10. Michels, A.; De Graaf, L. Examining citizen participation: Local participatory policy making and democracy. *Local Gov. Stud.* **2010**, *36*, 477–491. [CrossRef]
11. Granberg, M.; Åström, J. Planners support of e-participation in the field of urban planning. In *Handbook of Research on E-Planning: ICTs for Urban Development and Monitoring*; IGI Global: Hershey, PA, USA, 2010; pp. 237–251.
12. Barber, B. *Strong Democracy: Participatory Politics for a New Age*; University of California Press: Berkeley, CA, USA, 1984.
13. Macpherson, C.B. The Maximization of Democracy. In *Democratic Theory: Essays in Retrieval*; Macpherson, C.B., Ed.; Oxford University Press: Oxford, UK, 1973; pp. 3–24.
14. Pateman, C. *Participation and Democratic Theory*; Cambridge University Press: Cambridge, UK, 1970.
15. Refugee Data Finder. UNHCR. Available online: https://www.unhcr.org/refugee-statistics/ (accessed on 20 October 2020).
16. Statista. Immigration to Sweden. Available online: https://www.statista.com/statistics/523293/immigration-to-sweden// (accessed on 20 October 2020).
17. MSB [Swedish Contingencies Agency]. Om krisen eller kriget kommer. [If Crisis or War Comes-Important information for the population of Sweden]. 2018. Available online: https://www.msb.se/sv/publikationer/om-krisen-eller-kriget-kommer/ (accessed on 12 November 2020).
18. *Participatory Democracy and Political Participation: Can Participatory Engineering Bring Citizens Back in?* Zittel, T.; Fuchs, D. (Eds.) Routledge: Oxon, London, UK, 2007.
19. Blühdorn, I. *The Participatory Revolution: New Social Movements and Civil Society*; Wiley-Blackwell: Oxford, UK, 2009.
20. Karlsson, D. *Den Svenske Borgmästaren. Kommunstyrelsens Ordförande i Den Lokala Demokratin. [The Swedish Mayor. Chairman of the Municipal Executive Board in Local Democracy]*; Göteborgs Universitet: Förvaltningshögskolan, Sweden, 2006.
21. Martiniello, M. 4. Political Participation, Mobilisation and Representation of Immigrants and Their Offspring in Europe. In *Migration and Citizenship*; BauBöck, R., Ed.; Amsterdam University Press: Amsterdam, The Netherlands, 2006; pp. 83–105.

22. Goodman, S.W.; Wright, M. Does mandatory integration matter? Effects of civic requirements on immigrant socio-economic and political outcomes. *J. Ethn. Migr. Stud.* **2015**, *41*, 1885–1908. [CrossRef]
23. Smith, S.R.; Ingram, H. Policy tools and democracy. In *The Tools of Government: A Guide to the New Governance*; Oxford University Press: New York, NY, USA, 2002; pp. 565–584.
24. Clyne, M. The use of exclusionary language to manipulate opinion: John Howard, asylum seekers and the reemergence of political incorrectness in Australia. *J. Lang. Politics* **2005**, *4*, 173–196. [CrossRef]
25. Lane, P. Mediating national language management: The discourse of citizenship categorization in Norwegian media. *Lang. Policy* **2009**, *8*, 209–225. [CrossRef]
26. Barcena, E.; Read, T.; Sedano, B. An approximation to inclusive language in LMOOCs based on appraisal theory. *Open Linguistics* **2020**, *6*, 38–67. [CrossRef]
27. da Cruz, N.F.; Tavares, A.F.; Marques, R.C.; Jorge, S.; de Sousa, L. Measuring local government transparency. *Public Manag. Rev.* **2016**, *18*, 866–893. [CrossRef]
28. Mossberger, K.; Wu, Y.; Jimenez, B. *Can E—Government Promote Informed Citizenship and Civic Engagement? A Study of Local Government Websites in the US*; Midwest Political Science Association: Bloomington, Indiana, 2010.
29. OECD. *Digital Government Review of Sweden: Towards a Data-Driven Public Sector*; OECD Digital Government Studies; OECD Publishing: Paris, French, 2019; Available online: https://www.oecd.org/gov/digital-government/digital-government-review-of-sweden-4daf932b-en.htm (accessed on 22 January 2021).
30. Cumbie, B.A.; Kar, B. A study of local government website inclusiveness: The gap between e-government concept and practice. *Inf. Technol. Dev.* **2016**, *22*, 15–35. [CrossRef]
31. SALAR. *Information Till Alla. En Granskning av Kommunernas Webbplatser. [Information for Everyone. Review of Local Government Websites]*; SALAR: Stockholm, Sweden, 2017.
32. Bélanger, F.; Carter, L. The impact of the digital divide on e-government use. *Commun. ACM* **2009**, *52*, 132–135. [CrossRef]
33. Wæraas, A.; Bjørnå, H.; Moldenæs, T. Place, organization, democracy: Three strategies for municipal branding. *Public Manag. Rev.* **2015**, *17*, 1282–1304. [CrossRef]
34. Montin, S. *Politisk Styrning Och Demokrati i Kommunerna: Åtta Dilemman i Ett Historiskt Ljus. [Political Governance and Democracy in the Municipalities: Eight Dilemmas in a Historical Light]*; Sveriges Kommuner och Landsting: Stockholm, Sweden, 2006.
35. Johansson, F.; Nilsson, L.; Strömberg, L. *Kommunal Demokrati under Fyra Decennier. [Municipal Democracy for Four Decades]*; Liber: Malmö, Sweden, 2001.
36. Steyaert, J. Local governments online and the role of the resident: Government shop versus electronic community. *Soc. Sci. Comput. Rev.* **2000**, *18*, 3–16. [CrossRef]
37. Scott, J.K. "E" the people: Do US municipal government web sites support public involvement? *Public Adm. Rev.* **2006**, *66*, 341–353. [CrossRef]
38. Neves, J.P.; Felizes, J. E-participation in local governments: The case of Portugal. In *Towards DIY-Politics. Participatory and Direct Democracy at the Local Level in Europe*; Vanden Broele: Brugge, Belgien, 2007; pp. 277–295.
39. Pina, V.; Torres, L.; Royo, S. E-government evolution in EU local governments: A comparative perspective. *Online Inf. Rev.* **2009**, *33*, 1137–1168. [CrossRef]
40. Wirtz, B.W.; Piehler, R.; Daiser, P. E-government portal characteristics and individual appeal: An examination of e-government and citizen acceptance in the context of local administration portals. *J. Nonprofit Public Sect. Mark.* **2015**, *27*, 70–98. [CrossRef]
41. Sobaci, M.Z.; Eryigit, K.Y. Determinants of e-democracy adoption in Turkish municipalities: An analysis for spatial diffusion effect. *Local Gov. Stud.* **2015**, *41*, 445–469. [CrossRef]
42. Saglie, J.; Vabo, S.I. Size and e-democracy: Online participation in Norwegian local politics. *Scand. Political Stud.* **2009**, *32*, 382–401. [CrossRef]
43. Detlor, B.; Hupfer, M.E.; Ruhi, U.; Zhao, L. Information quality and community municipal portal use. *Gov. Inf. Q.* **2013**, *30*, 23–32. [CrossRef]
44. Lappas, G.; Triantafillidou, A.; Kleftodimos, A.; Yannas, P. Evaluation framework of local e-government and e-democracy: A citizens' perspective. In Proceedings of the 2015 IEEE Conference on e-Learning, e-Management and e-Services (IC3e), Melaka, Malaysia, 24–26 August 2015; pp. 181–186.
45. Simelio, N.; Ginesta, X.; de San Eugenio Vela, J.; Corcoy, M. Journalism, transparency and citizen participation: A methodological tool to evaluate information published on municipal websites. *Inf. Commun. Soc.* **2019**, *22*, 369–385. [CrossRef]
46. Guillamón, M.D.; Ríos, A.M.; Gesuele, B.; Metallo, C. Factors influencing social media use in local governments: The case of Italy and Spain. *Gov. Inf. Q.* **2016**, *33*, 460–471. [CrossRef]
47. Fashoro, I.; Barnard, L. Challenges to the successful implementation of social media in a South African municipality. In Proceedings of the South African Institute of Computer Scientists and Information Technologists; Association for Computing Machinery: New York, NY, USA, 2017; pp. 1–9.
48. Fashoro, I.; Barnard, L. Assessing South African Government's Use of Social Media for Citizen Participation. *Afr. J. Inf. Syst.* **2021**, *13*, 3.
49. Bonsón, E.; Royo, S.; Ratkai, M. Facebook practices in Western European municipalities: An empirical analysis of activity and citizens' engagement. *Adm. Soc.* **2017**, *49*, 320–347. [CrossRef]

50. Kardan, A.A.; Sadeghiani, A. Is e-government a way to e-democracy? A longitudinal study of the Iranian situation. *Gov. Inf. Q.* **2011**, *28*, 466–473. [CrossRef]
51. Lee, C.P.; Chang, K.; Berry, F.S. Testing the development and diffusion of e-government and e-democracy: A global perspective. *Public Adm. Rev.* **2011**, *71*, 444–454. [CrossRef]
52. Torpe, L.; Nielsen, J. Digital communication between local authorities and citizens in Denmark. *Local Gov. Stud.* **2004**, *30*, 230–244. [CrossRef]
53. Lindblad-Gidlund, K.; Giritli-Nygren, K. The Myth of E-Government. In *Handbook of Research on ICT-Enabled Transformational Government: A Global Perspective*; IGI Global: Hershey, PA, USA, 2009; pp. 313–328.
54. Lidén, G. Supply of and demand for e-democracy: A study of the Swedish case. *Inf. Polity* **2013**, *18*, 217–232. [CrossRef]
55. Klang, M.; Nolin, J. Disciplining social media: An analysis of social media policies in 26 Swedish municipalities. *First Monday* **2011**, *16*, 8. [CrossRef]
56. Larsson, A.O. Bringing it all back home? Social media practices by Swedish municipalities. *Eur. J. Commun.* **2013**, *28*, 681–695. [CrossRef]
57. Lidén, G.; Larsson, A.O. From 1.0 to 2.0: Swedish municipalities online. *J. Inf. Technol. Politics* **2016**, *13*, 339–351. [CrossRef]
58. Coleman, S.; Blumler, J.G. *The Internet and democratic citizenship: Theory, Practice and Policy*; Cambridge University Press: Cambridge, UK, 2009.
59. Marshall, T.H. *Citizenship and Social Class*; Pluto Press: Concord, MA, USA, 1992.
60. Stewart, A. Two conceptions of citizenship. *Br. J. Sociol.* **1995**, *46*, 63–78. [CrossRef]
61. Mossberger, K.; Tolbert, C.J.; McNeal, R.S. *Digital Citizenship: The Internet, Society, and Participation*; MIT Press: Cambridge, MA, USA, 2007.
62. Dobson, A. Environmental citizenship: Towards sustainable development. *Sustain. Dev.* **2007**, *15*, 276–285. [CrossRef]
63. Blank, Y. Spheres of citizenship. *Theor. Inq. Law* **2007**, *8*, 411–452. [CrossRef]
64. Cohen, J.L. Changing Paradigms of Citizenship and the Exclusiveness of the Demos. *Int. Sociol.* **1999**, *14*, 245–268. [CrossRef]
65. Carens, J.H. *Culture, Citizenship, and Community: A Contextual Exploration of Justice as Evenhandedness*; Oxford University Press on Demand: Oxford, UK, 2000.
66. Citizenship, Encyclopedia Britannica. Available online: https://www.britannica.com/topic/citizenship (accessed on 25 November 2020).
67. Joppke, C. Transformation of citizenship: Status, rights, identity. *Citizsh. Stud.* **2007**, *11*, 37–48. [CrossRef]
68. Baldi, G.; Goodman, S.W. Migrants into members: Social rights, civic requirements, and citizenship in Western Europe. *West. Eur. Politics* **2015**, *38*, 1152–1173. [CrossRef]
69. Aptekar, S. Citizenship and its Others. *Ethn. Racial Stud.* **2017**, *40*, 1345–1347. [CrossRef]
70. Dominelli, L.; Moosa-Mithal, M. Reconfiguring citizenship. Social Exclusion and Diversity within Inclusive Citizenship Practices. *J. Community Pract.* **2014**, *24*, 492–494.
71. Mouritsen, P. Beyond post-national citizenship. Access, consequence, conditionality. In *European Multiculturalism(s): Cultural, Religious and Ethnic Challenges*; Meer, N., Modood, T., Triandafyllidou, A., Eds.; Edinburgh University Press: London, UK, 2011; pp. 88–115.
72. Soysal, Y.N.; Soyland, A.J. *Limits of Citizenship: Migrants and Postnational Membership in Europe*; University of Chicago Press: Chicago, IL, USA, 1994.
73. Hammar, T. *Democracy and the Nation State: Aliens, Denizens and Citizens in a World of International Migration*; Aldershot: Avebury, UK, 1990.
74. Dingu-Kyrklund, E. *Citizenship, Migration, and Social Integration in Sweden: A Model for Europe?* CERIS Working Paper No. 52; Joint Centre of Excellence for Research on Immigration and Settlement: Toronto, ON, Canada, 2007.
75. Hansen, R. The poverty of postnationalism: Citizenship, immigration, and the new Europe. *Theory Soc.* **2009**, *38*, 1–24. [CrossRef]
76. Hagelund, A. After the refugee crisis: Public discourse and policy change in Denmark, Norway and Sweden. *Comp. Migr. Stud.* **2020**, *8*, 1–17. [CrossRef]
77. Aneesh, A.; Wolover, D.J. Citizenship and inequality in a global age. *Socio. Compass* **2017**, *11*, e12477. [CrossRef]
78. Milanovic, B. Global inequality recalculated and updated: The effect of new PPP estimates on global inequality and 2005 estimates. *J. Econ. Inequal.* **2012**, *10*, 1–18. [CrossRef]
79. Milanovic, B. Global income inequality in numbers: In history and now. *Glob. Policy* **2013**, *4*, 198–208. [CrossRef]
80. Balta, E.; Altan-Olcay, Ö. Strategic citizens of America: Transnational inequalities and transformation of citizenship. *Ethn. Racial Stud.* **2016**, *39*, 939–957. [CrossRef]
81. Harpaz, Y.; Mateos, P. Strategic citizenship: Negotiating membership in the age of dual nationality. *J. Ethn. Migr. Stud.* **2019**, *45*, 843–857. [CrossRef]
82. The World Bank. Available online: https://id4d.worldbank.org/global-dataset (accessed on 15 February 2021).
83. Open Society Foundation and Namati. A community-Based Practitioner's Guide. Documenting Citizenship & Other Forms of Legal Identity. 2018. Available online: https://www.justiceinitiative.org/uploads/286c1989-73db-4a17-b5a8-79706ccce5e4/a-community-based-practitioners-guide-documenting-citizenship-and-other-forms-of-legal-identity-20180627.pdf (accessed on 15 January 2021).
84. Bosniak, L. Varieties of citizenship. *Fordham L. Rev.* **2006**, *75*, 2449.

85. Ochoa Espejo, P. Taking place seriously: Territorial presence and the rights of immigrants. *J. Political Philos.* **2016**, *24*, 67–87. [CrossRef]
86. Strozzi, C. The changing nature of citizenship legislation. *IZA World Labor.* 2017, p. 322. Available online: https://wol.iza.org/uploads/articles/322/pdfs/changing-nature-of-citizenship-legislation.pdf (accessed on 15 January 2021).
87. Gallego, A. Where else does turnout decline come from? Education, age, generation and period effects in three European countries. *Scand. Political Stud.* **2009**, *32*, 23–44. [CrossRef]
88. Hooghe, M.; Kern, A. The tipping point between stability and decline: Trends in voter turnout, 1950–1980–2012. *Eur. Polit. Sci.* **2017**, *16*, 535–552. [CrossRef]
89. Bobbio, L. Designing effective public participation. *Policy and Society* **2019**, *38*, 41–57. [CrossRef]
90. Verba, S.; Schlozman, K.L.; Brady, H.E. *Voice and Equality: Civic Voluntarism in American Politics.*; Harvard University Press: Cambridge, MA, USA, 1995.
91. Arnstein, S.R. A ladder of citizen participation. *J. Am. Inst. Plan.* **1969**, *35*, 216–224. [CrossRef]
92. Gustafson, P.; Hertting, N. Understanding participatory governance: An analysis of participants' motives for participation. *Am. Rev. Public Adm.* **2017**, *47*, 538–549. [CrossRef]
93. Tahvilzadeh, N. Understanding participatory governance arrangements in urban politics: Idealist and cynical perspectives on the politics of citizen dialogues in Göteborg, Sweden. *Urban. Res. Pract.* **2015**, *8*, 238–254. [CrossRef]
94. Tahvilzadeh, N.; Kings, L. Under Pressure: Invited Participation amidst Planning Conflicts. In *Conflict in the City: Contested Urban Spaces and Local Democracy*; Gualini, E., Allegra, M., Mourato, J.M., Eds.; Jovis: Berlin, Germany, 2015; pp. 94–111.
95. Holgersson, J.; Karlsson, F. Public e-service development: Understanding citizens' conditions for participation. *Gov. Inf. Q.* **2014**, *31*, 396–410. [CrossRef]
96. Hertting, N.; Kugelberg, C. (Eds.) *Local Participatory Governance and Representative Democracy: Institutional Dilemmas in European Cities*; Routledge: London, UK, 2017.
97. Palumbo, A.; Bellamy, R. *Citizenship*; Routledge: London, UK, 2010.
98. Spång, M. *Svenskt Medborgarskap: Reglering och Förändring i ett Skandinaviskt Perspektiv. [Swedish Citizenship: Regulation and Change in a Scandinavian Perspective]*; Delegationen för Migrationsstudier (Delmi): Stockholm, Sweden, 2015.
99. Norwegian Citizenship. Available online: https://www.norden.org/en/info-norden/norwegian-citizenship (accessed on 22 March 2021).
100. SOU. *Kommunal Rösträtt för Invandrare. [Municipal Suffrage for Immigrants]*; Ministry of Local Government: Stockholm, Sweden, 1975.
101. SFS [Swedish Code Statutes]. *Vallagen*; Ministry of Justice: Stockholm, Sweden, 1997.
102. SOU. *Svenskt Medborgarskap. [Swedish Citizenship]*; Ministry of Justice: Stockholm, Sweden, 1999.
103. SOU. *Det svenska medborgarskapet. [Swedish Citizenship]*; Ministry of Justice: Stockholm, Sweden, 2013.
104. SOU. *Krav på kunskaper i svenska och samhällskunskap för svenskt medborgarskap. [Requirements or Knowledge of Swedish and Social Studies for Swedish Citizenship]*; Ministry of Justice: Stockholm, Sweden, 2021.
105. SCB, MIS. *Personer med Utländsk Bakgrund. Riktlinjer för Redovisning i Statistiken. [Statistics on Persons with Foreign Background Guidelines and Recommendations]*; SCB-Tryck: Örebro, Sweden, 2002. Available online: https://www.scb.se/contentassets/60768c27d88c434a8036d1fdb595bf65/mis-2002-3.pdf (accessed on 21 February 2021).
106. Swedish Citizenship. Available online: https://www.migrationsverket.se/Om-Migrationsverket/Statistik/Svenskt-medborgarskap.html (accessed on 11 March 2021).
107. The Swedish Local Government Act DS 2004:31. Available online: https://data.riksdagen.se/fil/9F02E1C3-CAC0-4940-9154-1E90A6425EBA (accessed on 16 December 2020).
108. Bevelander, P.; Spång, M. *Valdeltagande och Representation: Om Invandring och Politisk Integration i Sverige. [Electoral Participation and Representation: On Immigration and Political Integration in Sweden]*; Delmi rapport 2017:7; Delmi: Stockholm, Sweden, 2017.
109. Mäntysalo, R.; Saglie, I.L.; Cars, G. Between input legitimacy and output efficiency: Defensive routines and agonistic reflectivity in Nordic land-use planning. *Eur. Plan. Stud.* **2011**, *19*, 2109–2126. [CrossRef]
110. Amnå, E. *New Forms of Citizen Participation: Normative Implications*; Nomos Verlagsgesellschaft mbH Co. KG: Baden-Baden, Germany, 2010.
111. SOU. *En uthållig Demokrati! Politik för Folkstyrelse på 2000-Talet [Sustainable Democracy! Policy for Government by the People in the 2000s]*; Ministry of Justice: Stockholm, Sweden, 2000.
112. Proposition. In *Demokrati för Det Nya Seklet [Democracy for New Century]*; Ministry of Justice: Stockholm, Sweden, 2001.
113. Montin, S.; Granberg, M. *Moderna kommuner. [Modern Municipalities]*; Liber: Stockholm, Sweden, 2013.
114. SALAR. *Faktablad Projektet Medborgardialog 7 e-petioner. [Fact Sheet. Project Citizen Dialogue 6 e-Petions]*; SALAR: Stockholm, Sweden, 2009.
115. SOU. *Låt fler forma framtiden! [Let More People Form the Future!]*; Ministry of Culture: Stockholm, Sweden, 2016.
116. Guziana, B. Local democracy initiatives in Sweden: Inclusive or exclusive participatory democracy? In Advances in Sustainable Development Research, Proceedings of the 23rd International Sustainable Development Research Society Conference, Bogotá, Colombia, 14–16 June 2017.
117. Kekic, L. The Economist Intelligence Unit's index of democracy. *Economist* **2007**, *21*, 1–11.

118. Kaufmann, B. Sweden: Better late than never. Towards a stronger initiative right in local politics. In *Local Direct Democracy in Europe*; Schiller, T., Ed.; Verlag für Sozialwissenschaften: Wiesbaden, Deutschland, 2011; pp. 254–267.
119. Wallin, G. *Direkt demokrati: Det kommunala experimentalfältet. [Direct Democracy: The municipal experimental field]*; Statsvetenskapliga Institutionen: Stockholm, Sweden; Stockholms Universitet: Stockholm, Sweden, 2007.
120. SFS [Swedish Code Statutes]. Lag om kommunala folkomröstningar [Low om municipal referendums], 1994:692. Available online: https://www.riksdagen.se/sv/dokument-lagar/dokument/svensk-forfattningssamling/lag-1994692-om-kommunala-folkomrostningar_sfs-1994-692 (accessed on 15 March 2021).
121. SALAR. *Det Förstärkta Folkinitiativet. Erfarenheter Från Folkinitiativ Och Lokala Folkomröstningar 2011–2015. [The Strengthened People's Initiative. Experiences from People's Initiatives and Local Referendums 2011–2015]*; SALAR: Stockholm, Sweden, 2019.
122. SCB [Statistic Sweden]. Folkomröstningar. Available online: https://www.statistikdatabasen.scb.se/pxweb/sv/ssd/START__ME__ME0002__ME0002C/ME0002KnD01/?rxid=c6bc2ec4-e5ce-4c37-93ef-cda0bdf768af (accessed on 20 March 2020).
123. EMN [European Migration Network]. *Policy Report 2015 Sweden*; Swedish Migration Agency: Norrköping, Sweden, 2016.
124. Stenberg, J. (Ed.) *Framtiden är Redan Här: Hur Invånare kan bli Medskapare i Stadens Utveckling [The Future is Already Here. How Residents can Become Co-Creators in the City's Development]*; Chamers TekniskaHögskola: Göterborg, Sweden, 2013.
125. Abrahamsson, H.; Guevara, B.; Olofsson, G.; Svensson, J.; Tiger, A. *Invånardialogens Roll Och Former. Västra Götalandsregionens Samråd Med det Civila Samhället. [The Role and Forms of Resident Dialogue. The Västra Götaland Region's Consultation with Civil Society]*; KAIROS, Mistra Urban Futures Reports 2015; Mistra Urban Futures: Gothenburg, Sweden, 2015.
126. Abrahamsson, H. *Makt och Dialog i Rättvisa och Socialt Hållbara Svenska Städer. [Power and Dialogue in justice and socially sustainable Swedish cities]*; KAIROS 2013 Mistra Urban Futures: Gothenburg, Sweden, 2013.
127. Olofsson, G. *En Forskningsbaserad Essä Om Dialogens Möjligheter Och Hinder. [A Research-Based Essay on the Possibilities and Obstacles of Dialogue]*; Mistra Urban Futures Reports 2015; Mistra Urban Futures: Göteborg, Sweden, 2015.
128. PLAN. Planering och demokrati: Introduktion till specialnummer. [Planning and democracy: Introduction till special issues]. Available online: http://www.planering.org/plan-blog/2021/4/27/planering-och-demokrati-introduktion-till-specialnummer (accessed on 30 April 2020).
129. PLAN. Om PLAN. [About PLAN]. Available online: http://www.planering.org/om-plan (accessed on 30 April 2020).
130. Wiberg, S. *Medborgardialogens Politiska Organisering. [The Political Organization of Citizen Dialogue]*; Score: Stockholm, Sweden, Scores Rapportserie; 2016; Available online: http://www.decodeprojektet.se/media/1065/rapport-sofia-wiberg.pdf (accessed on 11 March 2021).
131. Jahnke, M.; Andersson, L.; Molnar, S. *Ledningssystem för Bättre Medborgardialog. [Management System for Better Citizen Dialogue]*; CMB, Kortrapport; CMB: Göteborg, Sweden, 2018; Available online: https://www.cmb-chalmers.se/wp-content/uploads/2018/10/Kortrapport_Diamap.pdf (accessed on 20 March 2021).
132. SALAR. *Underlag till Demokratibarometern. God Lokal Demokrati: En plattform. [Basis for the Democracy Barometer. Good Local Democracy: A Platform]*; SALAR: Stockholm, Sweden, 2011.
133. SALAR. *Det Reformerade Folkinitiativet Erfarenheter 2011–2013. [The Reformed people's Initiative Experiences 2011–2013]*; SALAR: Stockholm, Sweden, 2013.
134. Norris, P. (Ed.) *Critical Citizens: Global Support for Democratic Government*; OUP: Oxford, UK, 1999.
135. Mirchandani, D.A.; Hayes, J.P.; Kathawala, Y.A.; Chawla, S. Preferences of Kuwait's Residents for E-Government Services and Portal Factors. *J. Dev. Areas* **2018**, *52*, 269–279. [CrossRef]
136. DESA [United Nations Department of Economic and Social Affairs]. Creating an Inclusive Society. Practical Strategies to Promote Social Integration. 2009. Available online: https://www.un.org/esa/socdev/egms/docs/2009/Ghana/inclusive-society (accessed on ...).
137. Bloemraad, I. Theorizing the power of citizenship as claims-making. *J. Ethn. Migr. Stud.* **2018**, *44*, 4–26. [CrossRef]
138. De Rooij, E.A. Patterns of immigrant political participation: Explaining differences in types of political participation between immigrants and the majority population in Western Europe. *Eur. Sociol. Rev.* **2012**, *28*, 455–481. [CrossRef]
139. Pomatto, J. Deliberative Mini-Publics: The best is yet to come. *Representation* **2016**, *52*, 239–248. [CrossRef]
140. Ekman, J.; Amnå, E. Political participation and civic engagement: Towards a new typology. *Hum. Aff.* **2012**, *22*, 283–300. [CrossRef]

 sustainability

Article

Nuancing Holistic Simplicity in Sweden: A Statistical Exploration of Consumption, Age and Gender

Marco Eimermann [1,*], Urban Lindgren [2] and Linda Lundmark [3]

1 Department of Geography and Arctic Research Centre, Umeå University, 901 87 Umeå, Sweden
2 Department of Geography, Umeå University, 901 87 Umeå, Sweden; urban.lindgren@umu.se
3 Department of Geography and Centre for Regional Studies, Umeå University, 901 87 Umeå, Sweden; linda.lundmark@umu.se
* Correspondence: marco.eimermann@umu.se

Citation: Eimermann, M.; Lindgren, U.; Lundmark, L. Nuancing Holistic Simplicity in Sweden: A Statistical Exploration of Consumption, Age and Gender. *Sustainability* **2021**, *13*, 8340. https://doi.org/10.3390/su13158340

Academic Editor: Tan Yigitcanlar

Received: 12 May 2021
Accepted: 19 July 2021
Published: 26 July 2021

Publisher's Note: MDPI stays neutral with regard to jurisdictional claims in published maps and institutional affiliations.

Copyright: © 2021 by the authors. Licensee MDPI, Basel, Switzerland. This article is an open access article distributed under the terms and conditions of the Creative Commons Attribution (CC BY) license (https://creativecommons.org/licenses/by/4.0/).

Abstract: Studies of sustainable ways of life have hitherto made limited use of register data since, e.g., voluntary simplicity is usually identified through characteristics that cannot be found in data registers. Despite this, claims about these trends have been made in many countries, at times generalising the phenomena both in academia and media, based on anecdotal examples. This article draws on a quantifiable definition of holistic simplicity that includes certain fully measurable aspects, such as living in more affluent suburbs, moving to less affluent places and a significant reduction in individual work income. Other aspects are partially observable in register data, such as housing and car consumption. The advantage of this study is that it combines relevant theories around voluntary simplicity with register data that capture important characteristics of the entire national population (in this case, in Sweden) and thus, to some extent, also captures the magnitude of the phenomena. The article aims to statistically explore different demographic groups' probability of becoming holistic simplifiers in Sweden, regarding their consumption, gender and age. It discusses opportunities and limitations for advancing our knowledge on voluntary simplicity in Sweden, with current findings suggesting more of the same consumption patterns and only initial paths to degrowth. This is discussed in the context of individuals' agency in a state such as Sweden, which is changing from collectivist social democratic values to more neo-liberal conditions.

Keywords: consumption; degrowth; geography; register data; voluntary simplicity; Sweden

1. Introduction

Global change and sustainable development struggles have been on the political agenda for over 35 years, fuelled by arguments pro and contra growth-led development accompanied by technological innovations to solve growth-related challenges. In the early 1990s, scholars claimed that economic growth would lead to increased environmental sustainability as technological innovations reduce the negative impacts of growth [1]. In contrast, climate change debaters argue that global warming has been continuing, irrespective of technological development [2]. Many policy documents at different geographical levels have acknowledged the failure of the market forces to adhere to the initial and subsequent sustainable development goals. Rules and regulations intended to steer economic activity and diminish negative environmental and societal impacts of global economic activity have so far been insufficient.

At the grassroots level, societal movements indicate tendencies for bottom-up translocal transition initiatives [3] and 'green waves' of counter-urban migration [4–6]. This offers sustainable development opportunities inspired by degrowth to decrease excessive resource use [7]. Such tendencies can imply that geographical areas that have long fought population decline increasingly adopt right-sizing strategies of their population policies [8]. This article examines individual actions and manifest choices in the context of voluntary simplicity, which has seldom been linked with geographical studies in Sweden [9].

Early studies of current voluntary simplicity were conducted in Australia, North America and the UK [10–15]. Voluntary simplicity entails transitions such as voluntary career changes to gain more control over one's time [15,16] as part of a long-term strategy to increase one's quality of life, involving much less consumption and/or income than one's potential level [10]. This includes giving up "the compulsive purchase of material things that end up owning their owners" and stopping to sacrifice non-working activities for a job promotion [17] (p. 71). Voluntary simplicity reflects the growing interest in affluent societies to reduce overconsumption, to spend more time in activities that are in line with one's values and less time in the rat race [18].

Regarding the context of the state in international comparisons, living and working in Sweden has long been based on social-democratic values. Although the Swedish state is adopting more neoliberal policies, it differs from longer standing Anglo-Saxon neoliberal settings in which many previous studies were conducted. Individualisation has, e.g., progressed further in Anglo-Saxon "risk societies" [19] than in more collectivist societies such as Sweden. This has contributed to the emergence and growth of a precariat of unsafe individualised workers struggling with decreased well-being in Anglo-Saxon neoliberal economic acceleration [11,20–23]. Such acceleration is less palpable in Sweden, where workers' rights, including shorter workweeks, longer parental leaves, longer holidays and favourable retirement options have kept individual health- and wellbeing-related necessities for voluntary simplicity relatively low. In recent decades, however, Swedish state safety nets have been deregulated and pressure on individuals' responsibilities has increased while, at the same time, the moral plight for men and women to work full time has remained. The rationale has been that people should pay income taxes to contribute to the maintenance and equity of welfare state services and benefits [24].

Simultaneously, there has been an upswing in Swedish research and popular debate on voluntary simplicity and sustainable lifestyles as a possible counterforce to these pressures [6,25–29]. Different opinions exist of who the voluntary simplifiers are and whether or how they contribute to society. Some call voluntary simplicity a self-induced bottom-up form of 'luxurious communism' [30], in which affluent people work less to enjoy more time with each other and their hobbies of growing vegetables, driving water scooters and drinking wine. Others have pointed at gender and ethnic issues if the changed lifestyles at household level mean that mainly males continue their jobs and (white) females become home makers [13,14,31–33]. This in turn has implications at national levels, since sustainability transitions through working less imply paying lower income taxes, some argue that voluntary simplifiers do not contribute enough to sustaining the Swedish welfare state [29].

Nuancing some sensitivities in Swedish popular and academic debates, we focus on the consumption practices of holistic simplifiers, a sub-group of voluntary simplifiers defined as people who "adjust their whole life patterns according to the ethos of voluntary simplicity. They often move from affluent suburbs or gentrified parts of major cities to smaller towns, the countryside, farms and less affluent or urbanised parts of the country [...] with the explicit goal of leading a 'simpler' life" [10] (pp. 625–626).

Against this background, our study provides insights into holistic simplicity in Sweden. It aims to statistically explore different demographic groups' probability to become holistic simplifiers in Sweden, focussing on consumption, gender and age. We address two research questions regarding the consumption of housing and cars after a move away from affluent suburbs: (1) How (if at all) do the studied holistic simplifiers change their housing status? (2) How (if at all) do they change their car ownership in the household? We link these questions with financial costs and potential environmental effects.

Our data describe the demographic and socio-economic characteristics of holistic simplifiers in Sweden. A major contribution of this article is that it draws on quantitative register data, which is a novel method in this field. The data and related analyses can measure the relative magnitude of holistic simplifiers in a national population, i.e., how common the phenomenon is. We can also analyse the relationships between different individual factors in relation to holistic simplicity. This alleviates some challenges of drawing on smaller samples, which may not be representative of the whole population, may be based on a selection bias or may cause uncertainties regarding the strength of the relationships between factors and the significance of these relationships. The data are longitudinal, enabling us to follow anonymised individuals over time, to identify individuals who meet the requirements of being a holistic simplifier (reducing work income and moving away from more affluent urban areas), and to study whether they perform more comprehensive lifestyle changes. We address the research questions for a cohort selected from the data in the year 2014, which we follow until 2016 using the variables in Table 1.

Table 1. Variable descriptions. Variables measured in 2014 and regarding changes 2014–2016.

Variable	Description
Age group 30–34	Dummy variable = 1 if the individual is aged between 30 and 34
Age group 35–39	Dummy variable = 1 if the individual is aged between 35 and 39
Age group 40–44	Dummy variable = 1 if the individual is aged between 40 and 44
Age group 45–49	Dummy variable = 1 if the individual is aged between 45 and 49
Age group 50–54	Dummy variable = 1 if the individual is aged between 50 and 54
Age group 55–59	Dummy variable = 1 if the individual is aged between 55 and 59
Born in Sweden	Dummy variable = 1 if the individual was born in Sweden
Male	Dummy variable = 1 if the individual is male
University degree	Dummy variable = 1 if the individual obtained a University degree (three years or longer)
Unemployment	Dummy variable = 1 if the individual received unemp. benefits
Single	Dummy variable = 1 if the individual is single
Child in household	Dummy variable = 1 if household includes child(ren) under 16
Tenure: Renting	Dummy variable = 1 if the individual resides in rented apt.
Tenure: Condo	Dummy variable = 1 if the individual resides in condominium
Tenure: Ownership	Dummy variable = 1 if the individual resides in owned house
Renting to ownership	Dummy variable = 1 if the individual changed form of tenure between 2014 and 2016 from renting to ownership
Changes in size of residence	Number of square metres of residence 2016 minus number of square metres of residence 2014
Changes in residence's assessed value per m^2	Residence assessed value per square metre 2016 minus residence assessed value per square metre in 2014. Assessed value is calculated by the Tax Authority and attributed to all properties. It is assessed to be 75% of the market value.
Changes in car ownership	Number of cars owned by household 2016 minus number of cars owned by household 2014
Changes in car registration year	Registration year of household's newest car 2016 minus registration year of household's newest car 2014
Changes in household disposable income	Household disposable income 2016 minus household disposable income 2014 (SEK 100)

We present a literature review before explaining our methods and materials. We then present the results before discussing their implications. Showing that this entails 'more of the same' rather than 'new paths', we analyse urban–rural migration and discuss our results while linking them with potential sustainability practices using accessible language (alternating jargon with less complex terms). We round off with conclusions and future research regarding local capacity building in destination areas and degrowth.

2. Literature Review

Although it has been estimated that the number of annual working hours per worker has been declining since the 1870s e.g., [34], we here focus on more recent attention for work time reduction. Schor [12] related work time reduction with sustainable consumption in Western societies. Work time reduction can comprise various scenarios, such as "reduced average hours per job, average annual hours per person, [or] lower total hours per working life" [12] (p. 47). Opposing consumer-driven productivity growth in the global North [35] (p. 15), she urged for working less and stabilizing consumption. To allow all people to consume natural resources equally, Schor [12] saw more potential in changed consumer behaviour than in technological improvements. In other words, we may have more fuel-efficient cars, but "the rebound effect is that we are also driving more and buying more cars" [18] (p. 68). Human preferences adapt to levels of income, which makes averting income increases an efficient way to reduce Western consumerist lifestyles and save natural resources [12].

Some estimates suggest that as many as 200 million people are exploring a wide spectrum of 'simpler ways' of living in the West; that about 80% of voluntary simplifiers are based in urban centres, and that 22% sold or changed their car [18]. Of British adults aged 30 to 60, 25% had downshifted, equally representing different socio-demographic groups [14]. However, scholars are concerned with social inequality, as work time reduction can concentrate in specific income, age and gender groups [12,36]. Others note that, notwithstanding increased opportunities for part-time employment, voluntary simplicity has not become common in countries such as the UK or USA [37] despite alarms of overworked employees [38]. Although not conclusive, this indicates complex relationships between people's individual agency and wider structural contexts.

2.1. Conscious Consumption

Voluntary simplicity studies in Anglo-Saxon capitalist societies have considered the importance of material wellbeing, personal consumption and quality of life [10,18]. The latter is connected with post-materialist values, such as the desire for more freedom, a stronger sense of community and more influence in democratic processes [39]. In this context, Etzioni [10] (p. 620) describes voluntary simplicity as "the choice out of free will [...] to limit expenditures on consumer goods and services, and to cultivate non-materialistic sources of satisfaction and meaning". He noted that even highly dedicated voluntary simplifiers pursue combinations of a reasonable level of work and consumption to attend to basic needs, with satisfaction derived from knowledge rather than consumer objects [10] (p. 637). Etzioni [10] described different levels of intensity leading to three variations of voluntary simplicity seekers: downshifters as a moderate form, strong simplifiers who give up high levels of income and socio-economic status, and holistic simplifiers. We operationalize the latter variation below, which we selected because its definition includes migration (see this article's introduction).

One caveat is that consumption is a complex issue, as, e.g., postmaterialist values in the USA doubled between 1972 and 1991 while personal consumption continued to grow [10] (p. 620). From a psychological perspective, the visibility of consumer goods is one pivotal aspect in traditional capitalist terms: displaying one's income by buying expensive-status goods signals success [10,20]. In such contexts, there are few established means to signal that one has opted for simplicity willingly rather than by necessity. This may, however, be achieved by using select consumer goods that are clearly associated with a simpler life pattern and that are as visible as traditional status symbols, such as a dressing down, but still wear some expensive items [10]. This "conspicuous non-consumption" [40] can also involve cooking at home more often instead of eating out, giving up expensive holidays [41], or buying less complicated and more modest rural houses. These signals change over time and vary between subcultures [10].

Moreover, recent studies highlight that voluntary simplifiers may own significantly fewer but more expensive and environmentally friendly durable consumer goods [42]. Thus, they may not necessarily spend less money on these goods, since their quality and price can be higher [43]. However, due to a lack of comparable data, it is difficult to make assessments regarding the extent of this phenomenon in different countries and contexts.

Although voluntary simplicity is about reduced consumption, it does not necessarily mean sacrificing comfort or enduring hardship [37,41]. Ragusa's [37] economic study of domestic urban–rural migrants in Australia indicated that few desired to fundamentally change their standard or way of living. Almost none of the 53 (mostly middle aged) respondents rejected contemporary consumerism, and all continued to drive cars, often commuting even longer distances. Several of them moved to increase their purchasing power, e.g., on the housing market while still pursuing economically rewarding careers. None made significant changes to counter dominant Western lifestyle trends, which pointed

2.2. Gender and Age

Results on gender and voluntary simplicity are inconsistent. Although more females in the survey of Kennedy et al. [44] linked downshifting with a desire to stay at home with children, the authors found no significant interaction effects between voluntary simplicity and gender. It may be common that men and women face different issues due to internalised gender roles relating to work [13], and that women experience more stress in combining paid work with unpaid household tasks [32]. Tan [13] found that women experienced more difficulty with career transitions than men did. Explanations for this could be that more women in her study lived without a partner, that they were in early phases of their transition or that they underwent changes more rapidly. Further, a number of less egalitarian men disliked the consequences of having less money or having to rely more on their female partner's income [13]. In Hamilton's [14] study, it was more common for women to stop working at all and more common for men to work less. Other studies contradict each other, e.g., Grigsby [45] found that women were more likely to be voluntary simplifiers while Hamilton & Denniss [46] concluded the opposite. Hence, Kennedy et al. [44] signal a need for more nuanced future research regarding gender and voluntary simplicity, arguing that structural changes in broader domains such as work culture and support for families would be required to increase sustainable practices. This further indicates that the context of the nation as ordering and structuring everyday lives needs to be included in the assessment of who is engaging in voluntary simplicity and what the motivations might be.

In particular, a welfare state such as Sweden promotes gender equality, but full gender equality encompasses more than men and women working equally and earning equal incomes [47]. The Swedish model's narrow interpretation of women's emancipation has contributed to a situation in which a sharp line exists between market wages for the breadwinner in the male-dominated public spheres and unpaid (care) work often carried out by women in domestic spheres. As the Swedish welfare state applies social security rights based on work performance and labour market participation, many women struggle with socio-economic lag [48] and a double workload of both work and unpaid house work. This has led to unequal power relations where the social and economic positions of males are often stronger than those of females, and women are over-represented in statistics on mental illness and on sick leave [49]. Time spent on vital reproductive work, such as caring for partners, children and relatives is often made invisible.

Time use is also linked with age, life course stage and life events such as child rearing and mortality [13,44]. Age and life course stage were significant in a Canadian study [44], as older respondents and those living in central urban neighbourhoods were more likely to be satisfied with their time use than those with children living at home. Tan's [13] study focused on the midlife stage, which involves considerable responsibilities in terms of family and financial commitments, while also being at the peak of career earning potential. Tan [13] indicated a changed relationship with time, as many studied individuals started to reach an age that signalled mortality. This made them realise that less time was available for establishment in alternative careers and other directions. On a structural level, contemporary midlife Australians' life plans were dominated by insecurity about comfortable retirement, as the Federal government had made it clear that they could no longer rely on state pensions to meet their needs [41].

2.3. Urban–Rural Relationships

In a Canadian study, respondents who had shifted to lower income were more likely to engage in sustainable household practises, such as reduced consumption [44]. The regression models demonstrated significant effects on the subjective well-being of owning one's home, but also that the decision to earn less did not appear to change patterns of car use [44]. They stated that environmental benefits of voluntary simplicity did not extend beyond the household level because pro-environmental behaviour, such as reduced car use, requires systemic changes in spatial planning. Although voluntary simplicity, as such, was not related to the place of residence (e.g., central urban, suburban), neighbourhood of residence was the strongest predictor of sustainable transportation practices [44]. Municipal spending in Canadian suburbs (where potential simplifiers live) had shifted away from public transportation in favour of multiple-car households. In this North American context, suburbs force reliance on cars and contribute to consumerist lifestyles [50]. This is a missed opportunity, since Alexander and Ussher [18] suggested that many simplifiers wish to escape the car culture but, for various reasons (e.g., harsh winters, health conditions or limited public transport), found this difficult or impossible.

In general, urban–rural relationships refer to functional linkages and interactions between urban and rural areas. They cover a spectrum of interactions through housing, employment, education, transport, tourism and resource use, including social transactions, administrative and service provision, and the movement of people, goods, and capital [51]. Within this spectrum, we here focus on socio-economic forces from economic geographic and voluntary simplicity perspectives. In economic geography, studies have addressed the ongoing urban concentration of companies, capital and individuals after the end of the Fordist production system, leading to uneven geographies of labour and growth [52,53]. These processes have resulted in increased levels of unemployment and out-migration from peripheral areas previously dominated by a single, large-scale workplaces [54]. However, Swedish studies have highlighted comparative advantages and successful companies in rural areas, as well as new ways of working and living there [55]. Where one lives and works is very important, as location is essential in the global competition for labour and capital [54]. Hence, Lindgren et al. [55] argue for a re-evaluation of the urban–rural divide.

Such a re-evaluation here includes holistic simplifiers' urban–rural moves that may be motivated by wishes to work and consume less. Voluntary simplifiers may contribute to local transformative capacity in Swedish rural areas by their social and human capital (e.g., work engagement, networks, skills [37]). For instance, Sandow and Lundholm [5] found a small but steady outflow of highly educated adults and their families from Swedish metropolitan areas to medium-sized and small towns (in 2003–2013). Most of them were public sector professionals or males working within arts and crafts [5]. In another example, retirees' spontaneous activism for the protection and reclamation of a riverbank in West Sussex (UK) was studied in the context of "nowtopias" [56]. These are territorial processes of regeneration that involve non-wage labour and are motivated by a desire to produce alternative local futures here and now, e.g., through the everyday experimentation of other worlds [57,58]. This provides a context for our urban–rural study below.

Our study thus employs a relational approach to holistic simplicity-induced urban–rural migration, which can link urban and rural areas with each other. Though starting from similar socio-economic globalisation processes, such as growth and degrowth, a relational approach recognises that globalisation affects people and settlements differently [59]. The unfolding population geographies in different places can be viewed as historically contingent developments, along with the evolving nature of translocal relations between urban and rural settlements [60,61]. Relational approaches shed light on irregular expansion and contraction of settlements and the acceleration or deceleration of local and regional linkages on a daily basis and over a life course [62,63]. Thus, urban–rural relationships shaping transformative capacity and nowtopias provide the context for our conclusions.

3. Materials and Methods

Most recent Swedish voluntary simplicity studies draw on deep knowledge gained from qualitative data of small samples in terms of socio-economic and geographic factors (e.g., [6,26,27]). Here, in contrast, we draw on rich georeferenced register data to expand the scope while exploring variables both on the individual and household levels. The empirical analysis is based on relevant longitudinal demographic and socioeconomic attributes of individuals in the whole country. These data have been created by matching a number of administrative registers at Statistics Sweden (SCB). No ethical approval was needed, since the personal information was not sensitive and the key between persons and codes was destroyed. The entire data set covers the period 1985 to 2016 and includes annual information about all Swedish inhabitants (9.99 mln in 2016: 4.98 mln women and 5.01 mln men. [64]). The database contains over 100 unidentified individual attributes annually. These attributes refer to demography, household (e.g., partners with their attributes, children), education, employment, unemployment (e.g., unemployment benefits), income and transfer payments (e.g., income from work, disposable income, pensions), housing characteristics (where registered; rented flat, condominium, detached house, etc.), coordinates of places of residence, etc.

As the data are based on official registers, the quality of the information is generally very high. Another advantage is that this type of data covers the entire population. This means that all individuals can be analysed, and there is no need to consider how to deal with nonresponse as is commonly the case in studies based on surveys. The richness of information provides a possibility to learn more about the magnitude, characteristics and distribution of this phenomenon.

We recognize that such data cannot capture people's entire consumption behaviour, let alone wellbeing or experienced changes in overall quality of life. Still, this study provides a good opportunity to operationalize Etzioni's [10] theorization of holistic simplifiers, i.e., identify individuals who live in affluent city neighbourhoods, substantially decrease their work income and move elsewhere during this process.

This offers a possibility to study to what extent these movers make changes towards a simpler life regarding two commonly considered major household expenditures (housing and car ownership), which may cause decreased general wellbeing through higher stress levels regarding household economy. The longitudinal qualities of register data enable us to follow individuals over time and observe whether they actually led a simpler life after they moved away from the affluent neighbourhood and decreased work income. Table 1 presents the variables in the analysis.

Steps in the Analysis

To identify holistic simplifiers, data needed to be organized in a number of steps. Step 1. We identified individuals aged 30–59 in 2014 and who were still alive in 2015 or 2016. Individuals younger than 30 were excluded since many of their residential moves and changes in level of consumption may be related to education (a common feature in Sweden), which could distort the analysis. Further, a shift to a simpler life entails an initial income level and relevant working experience to shift down from. Both may be lacking to a larger degree among young adults under 30. Studying this age group's potential motivations for living simpler lives would be more suitable for qualitative investigations.

Step 2. We selected those aged 30–59 who lived in one of Sweden's major cities (Stockholm, Gothenburg, Malmö), or in a regional capital city (in order of population size from larger to smaller: Uppsala, Linköping, Västerås, Örebro, Helsingborg, Norrköping, Jönköping, Umeå, Lund, Sundsvall, Karlstad, Växjö, Luleå, Östersund) in 2014, in a parish whose average income from work exceeds the city average income. We used this as a proxy for Etzioni's [10] 'affluent suburbs'.

Step 3. We selected holistic simplifiers as those who had moved in 2015 or 2016 to other places in Sweden (outside the selected affluent suburbs), and who had reduced their income from work with at least 50% after the move. We acknowledge that a reduction of 50% may seem like much, and previous studies have settled for smaller reductions. However, our motivation for using the reduction of 50% is that we can be quite certain these individuals have good opportunities to lead a substantially simpler life than before reducing their incomes. We control for forced unemployment in this study through a variable indicating unemployment benefits. We can thus be confident that the results of the other variables do not depend on unemployment. Although unemployment in this study does have a positive effect on the probability of consuming less, this effect is separate from the other results. We thus used this as a proxy for Etzioni's [10] 'leading a simpler life'.

Step 4. The reference group then consisted of individuals aged 30–59 living in the selected parishes in 2014, and who were alive in 2016, but had not moved to other parishes with lower average income. This may mean they had not moved at all, or that they had moved to another affluent parish with the same (or higher) average income from work.

These steps indicated about 1.4 million registered residential moves within Sweden, undertaken by 1.21 mln individuals (of whom some moved more than once). By using the criteria above, we identified 3188 individuals as holistic simplifiers. On an annual basis this corresponds to 0.11% of all movers. Some holistic simplifiers move longer distances, which justifies a comparison with a sub-selection of long-distance movers. The number of movers across municipality borders (290 municipalities in total) amounted to 488,000 the same year, which indicates that holistic simplifiers were 0.33% of all movers across municipality borders on an annual basis.

The modelling of Etzioni's [10] theorization of holistic simplifiers is carried out by estimating a binomial logit model capturing the differences between holistic simplifiers and the reference group. Since we are interested in comparing these two groups we constructed a dichotomous dependent variable, making, e.g., linear regression models (OLS) less useful. The applied software is Microsoft SQL Server (a database engine) and SPSS. The logit model applies a logistic function to model the binary dependent variable. On the right-hand side of the equation there is a vector (X) of observable attributes of the individual and a vector of parameters (beta) to be estimated. The variables and parameter estimates are presented in Table 2.

Table 2. Estimates of holistic simplifiers.

Variables	B	Sig.	Exp(B)
Individual level (in 2014)			
Age group 30–34	ref.		
Age group 35–39	0.089	0.394	0.915
Age group 40–44	−0.277	0.034	0.758
Age group 45–49	−0.517	0.000	0.596
Age group 50–54	−0.198	0.137	0.820
Age group 55–59	−0.070	0.601	0.933
Born in Sweden	0.161	0.106	1.174
Male	0.056	0.457	1.058
University degree	−0.689	0.029	0.502
Unemployment	0.543	0.000	1.721
Household level (in 2014)			
Single	0.367	0.000	1.443
Child in household	−0.450	0.000	0.638
Tenure: Renting	−0.271	0.006	0.763
Tenure: Condo	ref.		
Tenure: Ownership	−0.404	0.000	0.668
Changes 2014–2016			
Renting to ownership	2.117	0.000	8.308
Changes in size of residence	0.019	0.000	1.019
Changes in residence ass. value/m^2	−0.002	0.188	0.998
Changes in car ownership	0.129	0.192	1.137
Changes in car registration year	−0.029	0.004	0.972
Changes in household disp. income	0.000	0.002	1.000
Constant	−5.373	0.000	0.005
N = 589,301. Log-L = −10,191			

The model estimates in the results section are based on a population aged 30 to 59, which more or less excludes retirees. We chose to do so since we are interested in what holistic simplicity looks like among people at working age. Retirees reduce work income by definition, and they can be seen as downshifters as a consequence of age rather than own choice, but the philosophical question to what extent pre-retirees are voluntary simplifiers is beyond the scope of this article. To reduce the risk of biased results due to family changes during the study period (e.g., family dissolution or family formation) we only included holistic simplifiers who had not separated, changed partner or found a partner. As a consequence, changes in, for example, household disposable income, cannot be attributed to the income of a new spouse.

4. Results

4.1. Basic Demographic and Socio-Economic Characteristics of Holistic Simplifiers in Sweden

Table 2 presents model estimates of holistic simplifiers. First, we present and comment on individual demographic and socioeconomic variables, before shifting to household- and housing-related variables. To begin with, the probability of being a holistic simplifier varies by age. It is less likely that people in their forties make this type of change in their lives. This negative effect fades away as people grow older. People in their fifties are as likely as the reference category of people in their early thirties to engage in holistic simplicity. It can be added that we have estimated the same model including elderly people. These results showed the same patterns across age groups as presented in Table 2, but with the addition that the estimates turn significantly positive for all age groups over 60 except for the age group 75–79. These results suggest, when controlling for a wide array of demographic and socioeconomic factors, that people in their forties are the less likely age group to become holistic simplifiers.

The analysis shows that there are no significant differences between people born in Sweden and people born abroad, or between men and women.

Furthermore, people with a university degree are much less likely to be a holistic simplifier. The odds ratio indicates a drop by 50%. Being eligible for unemployment benefits seems to have a positive effect on the probability to be a holistic simplifier. The odds ratio increases by 72%. Taken together, these results show that Swedish holistic simplifiers are likely to have shorter formal education and more unemployment experience. These findings do not support the idea that opting for simplicity is a high-status phenomenon.

At the household level, the results indicate that family situation is important. In comparison to couples, people who are single are much more likely to be holistic simplifiers. The odds ratio increases by 44%. However, having underaged children in the household decreases the probability of making a change along these lines.

Further, the results show that, in comparison to living in condos, both living in rental flats and owned houses means a lower likelihood of becoming a holistic simplifier. On the other hand, it is equally correct to state that people living in condos are more likely than renters and house owners to become holistic simplifiers. Since people living in condos are usually regarded as owning their housing, this becomes a little complicated. Technically, condo owners do not own a property, they own a membership in a housing cooperative, which owns the property. True ownership of flats is a rare tenure in Sweden. We can only speculate about possible explanations, but households that have been able to purchase a house in affluent suburbs may have fewer reasons to move, not least because of the amenities such properties bring (e.g., own garden, tranquil surroundings, less traffic, etc.). People living in rented apartments may lack financial means to make a change in life in line with our interpretation of holistic simplification, partly because such moves usually involve buying a house. Access to rented housing is sometimes scarce in the destinations.

4.2. How (If at All) Do Holistic Simplifiers Change Their Housing Status?

Drawing on longitudinal data, we observe what happened to the holistic simplifiers after they moved out of the affluent suburbs and decreased their work income. This is operationalized by creating a number of variables showing differences between 2014 and 2016. The first variable in this category is changing tenure from renting to ownership. This estimate is positive and highly significant, implying that holistic simplifiers tend to shift from renting to owning their residences. As noticed above, this is a small group of people, but those who perform these changes are very likely to end up having their own property. One explanation for these results could be the ownership of second homes in the countryside, which is more widespread in Sweden than, for example, in Anglo-Saxon countries [65]. People may have moved to their second homes, which they usually own. In contrast to the imaginaries in 'luxurious communism' [30], this does not have to be a luxurious house. Rather, as part of the process towards holistic simplicity, the surrounding place may become more important [27].

To study housing situations more thoroughly, we constructed another variable regarding changes in the size of housing space. This may be a relevant factor because housing costs are commonly a large part of household expenditures. The estimate of this variable is positive, which means that an increase in living space increases the likelihood of being a holistic simplifier. The odds of being a holistic simplifier rise by 1.9% for each square metre of increased housing space. Thus, holistic simplifiers do not seem to reduce living space to any large extent, which is somewhat surprising, since leading a simpler life is amongst the other factors associated with less housing consumption.

It can, however, be argued that houses are cheaper far away from affluent suburbs where people get the chance of owning a bigger house at a lower cost. In order to control for changes in living space, we created a variable that measures differences in assessed value per square metre. This estimate turned out to be insignificant, implying that holistic simplifiers do not reduce their housing consumption after the move—they tend to have houses that are as expensive as those of the non-holistic simplifiers who stayed put in affluent suburbs. This is a surprising result, considering the spirit of holistic simplicity.

4.3. How (If at All) Do Holistic Simplifiers Change Their Consumption of Cars in the Household?

Another indicator of conscious consumption is car ownership [10,18]. This also represents differences of distance to commercial and welfare services, since people living farther away from dense urban areas often need more cars for their (daily) transports. The variable "changes in car ownership" measures the difference in the number of owned cars. The variable shows an insignificant estimate, indicating that holistic simplifiers do not reduce their car fleet, which seems to be contrary to expectation. Cars require money to buy and maintain and might, therefore, be a good candidate for cost reduction when planning for a simpler life. On the other hand, living farther away from dense urban areas, sometimes even in the countryside, may be difficult without a car. Public transport is scarce and schools, leisure activities and service facilities may be far off, making it more or less impossible to manage without a car.

Car expenditures are not only related to the number of cars the household owns. These expenditures are also connected to the age of the car. Generally, newer cars are more expensive than older cars due to, for example, capital costs and value depreciation. We created a variable measuring differences in age of a household's newest car before and after the move. This variable is negatively significant (Table 2), indicating that holistic simplifiers own older cars than the reference group. An increase of one year (i.e., a year's newer car model) decreases the likelihood of being a holistic simplifier by 2.8%. These holistic simplifiers thus cut car costs, to some extent, which was expected [10,18], but they do not reduce the number of cars, which is contrary to the radical lifestyle changes suggested in other studies [66].

The results so far suggest a limited reduction in car consumption and no reduction in consumption related to housing, but we know, by definition, that individuals meeting the holistic simplifier criteria in this study have reduced their work income by at least 50%. This income reduction suggests diminished consumption somewhere in the household budget. Households may reduce spending on newer cars, and other products and services we cannot observe. Alternatively, the households do not reduce their consumption at all, because they have access to other sources of income. To investigate this, we created a variable showing differences in household disposable income between 2014 and 2016. This includes all sources of income (work income, capital income, transfer payments, social benefits, etc.) for all individuals in the household (the holistic simplifier, his/her partner, income earned by youth living at home, etc.). The estimate of this variable reveals a positive significant effect on the probability of being a holistic simplifier, meaning that holistic simplifiers are more likely to have increased their household disposable income than their former affluent suburb neighbours. This is not what we expected from holistic simplifiers who supposedly shift to a simpler life. The reasons for this result are beyond the scope of this study but are well worth investigating in future research.

All in all, the results suggest that only a few of the 3,188 identified holistic simplifiers in Swedish register data reduced their consumption in accordance with what could be expected from previous studies. In fact, looking closer into data conditioning on reduced housing consumption and decreased household disposable income would reveal an even lower number of "true" holistic simplifiers [10].

5. Discussion

This article on Swedish conditions revealed two rather unique potentials for simplicity studies: (1) studying living and working conditions in egalitarian Nordic welfare states in comparison to previous studies in Anglo-Saxon contexts [11,12,20–23], and (2) drawing on rich (Swedish) longitudinal georeferenced register data. The former has implications for housing markets and labour markets, while the latter indicates potential for future studies.

Four reflections further explain this study's results and limitations. First, our results may diverge from findings elsewhere, as Sweden might not be comparable to other countries and contexts [18,44]. Using Etzioni's [10] criteria, we identified 3188 persons as holistic simplifiers in Sweden during the period 2014–2016. Such a low number indicates that this phenomenon is marginal at best, using this definition. It is less likely than in other studies [41,44] that people in their forties make this type of lifestyle change. We found no significant differences between people born in Sweden and elsewhere, or between men and women (which differs from [13,14,45]. Other characteristics are more important for increasing the likelihood to become a holistic simplifier, such as shorter formal education and more unemployment experience. Singles are much more likely than couples to be holistic simplifiers. As expected, having underaged children in the household decreases the probability of making such lifestyle changes.

The following question arises: Do similar patterns emerge in other countries when drawing on register data and following the same operationalisation as in this study? It is plausible that our operationalisation of Etzioni's [10] definition of holistic simplifiers differs from the original intentions, and that it does not fully capture the phenomenon as intended. We conducted our study 20 years later and in a different country with different structural contexts. Differences in a society's social and political structure, as well as general living conditions, probably impact the likelihood of becoming a holistic simplifier, but this is an under-researched field.

Second, this relates with ideas that the welfare state allows for certain degrees of voluntary simplicity without a need to resign from employment, find less expensive housing or relocate [28]. This is also linked with life course stages as, e.g., parents are entitled to parental leave with only little loss of income from work (paid for by all citizens' work income taxes [67]). This means that there is no essential income reduction when at home taking care of children, which may leave room for temporary voluntary simplicity in terms of reduced workload both for women and men [48]. This regards the age groups in which other studies suggest high incidence of voluntary simplicity [13,41,44]. We thus find that the welfare state's social benefit schemes need more attention when studying how simplifiers' individual choices and practices are influenced by benefits such as longer parental leave or sick leave (e.g., [49]). Welfare states such as Sweden—while in a process of neoliberal changes—may keep individual health- and wellbeing-related necessities for voluntary simplicity relatively low.

A third reflection regards the extent to which individuals actually engage in holistic simplicity, voluntary simplicity or downshifting. Although these are similar concepts, the individuals studied here may engage in some convenient changes without fully embracing overall lifestyle shifts [37]. Differences between qualitative and quantitative samples are underexplored: are interviewees in other studies a (more hardcore) subgroup of this article's data set, or can this study's 50% income reduction criterion indicate a minor subgroup of potential holistic simplifiers? More thorough studies facilitate further conceptual reflections on sustainable living through reduced worktime and consumption.

Fourth, difficulties in terms of operationalisation of Etzioni's [10] description of holistic simplifiers and related theories of conscious consumption [40,41] imply that we cannot measure all complex facets of consumption. Many consumption indicators are hidden in the register data, as, e.g., costs of leisure activities and social practices may be significantly lower in less affluent areas. The studied individuals can be living simpler lives through consuming less clothes and gadgets or eating out less often [40]. If this is the case, and since their disposable household income is not reduced, they may save much money. Further research could study potential alternative consumption issues; is the disposable household income invested in any way, spent on business ventures, bestowed on charity, or do other strategies exist?

A limitation of this study is that we do not know the studied households' reasons for not purchasing newer (possibly more environmentally friendly) consumer goods such as cars (e.g., [18,42]), which can also be investigated in future research. More in-depth overviews indicate how human values may guide voluntary simplicity lifestyles [43,68].

Future Studies Drawing on Georeferenced Longitudinal Register Data

Linking individual agency with geographic and political structures, future studies in Nordic contexts can reveal concrete policies and practices (e.g., regarding unemployment benefits and parental leaves) that differ from the hitherto studied Anglo-Saxon contexts. Such benefits and other features can be studied while considering interplays between individual lifestyle choices, trends and norms in society and states' structural components related with political ideologies. This links voluntary simplicity more profoundly with individualisation processes [19] and economic acceleration impacts on Nordic citizens' everyday lives [11,20–22].

Although this study's register data are rich in regard to the studied number of individuals and characteristics, the short time span (2014–2016) merely provides a snapshot of patterns and statistical relationships during the mid-2010s, with limited possibilities to measure evolving intentions and motivations. Future studies could utilize the longitudinal qualities of register data to better explore the dynamics of voluntary simplicity within a country's whole population. We suggest two different approaches.

One approach follows the evidence in the literature that lifestyles such as voluntary simplicity have been on people's minds for quite some time. Macro-economic studies indicate ongoing work-hour decreases in most industrialized countries [12,34]. To study whether this trend is as visible on the micro level of individuals and their specific life situations, this article's research design can be combined with longitudinal register data going back to the early 1990s. Individual- and household-level variable estimates could be compared over longer periods of time by stepwise repeating our analysis for moving three-year time frames for the periods 1990–1992, 1991–1993, etc. until 2018–2020, when available. This may reveal voluntary simplicity trends and provide information about changing numbers of individuals engaging in holistic simplicity.

Another approach considers that voluntary simplicity is not likely to occur overnight. In this study, we observe individuals over a three-year period, which may be too narrow a time frame. Longitudinal data can observe individual actions and events over a much longer period of time. Future studies could analyse the extent to which people make gradual shifts that, consciously or not, steer towards lifestyle changes. For example, a second home purchased one or two decades ago—which from the start was entirely considered to be used for recreational purposes during summer—could gradually be perceived to have qualities that make people think about voluntary simplicity [65]. Evidence suggests that second home owners of all ages use their second homes as a way to experience quality of life without urban demands and burdens, to relax and for practices such as gardening, outdoor recreation and house work, which they do not have time for during a regular work week [69]. In this process, the second home may become a primary residence, which could reduce household expenditures and enable a livelihood that is less dependent on work income. Drawing on register data with a biographic point of view can reveal the life paths of voluntary simplifiers over longer periods.

These insights indicate the need to link studies of regular residences and second homes with voluntary simplicity, as future research can investigate extents to which voluntary simplifiers have access to second homes (compared to others), locations of these homes and arising gender issues [13,32]. Analysis focusing on sequences of events in the life course can identify different ways towards a simpler life, including the length of such a process or distinctions between more and less common ways to simplicity (including car ownership and use). This is a relevant policy area for Swedish and similar municipalities with relatively many second home residents. Their contributions to transformative capacity building [56] through social and human capital (e.g., skills, work engagement, networks [37]) is relevant for right-sizing strategies in rural municipalities' planning of housing and infrastructure for (electric) cars and other means of transport [8].

6. Conclusions

This article aimed to statistically explore different demographic groups' probability of becoming holistic simplifiers in Sweden, focussing on consumption, gender and age. It showed complex and paradoxical relationships between property ownership and likelihood of becoming a holistic simplifier. The exact reasons are unknown, but holistic simplifiers tend to shift to owning their residences. Larger living spaces increase the likelihood of being a holistic simplifier, and the assessed value per square meter does not imply a reduction in holistic simplifiers' housing size after the move either. They do not reduce the number of owned cars, but they tend to own somewhat older cars than the reference group. Finally, holistic simplifiers are more likely to have increased their household disposable income than their former neighbours who stayed in affluent suburbs.

These unexpected findings differ from previous studies elsewhere (e.g., [66]). Contrary to previous studies [11–15] and popular discourse of voluntary simplicity, our data show that holistic simplicity as identified here is, or at least until recently has been, a marginal phenomenon in Sweden. The group of individuals identified as holistic simplifiers in the data is a small part of the total population, and those performing lifestyle changes in accordance with our informed expectations are even fewer.

At first sight, our study thus suggests 'more of the same', as the studied holistic simplifiers undertook urban–rural moves without drastically reducing their consumption of housing and cars (in line with [5]). Their move from affluent suburbs to near and remote countryside nevertheless signals novel contexts that plea for re-evaluating the urban–rural divide [55]. Such studies could consider geographic and social inequalities and translocal relations between urban and rural settlements [59–61]. As indicated by relational approaches, there is no blueprint for degrowth through voluntary simplicity, but this is rather a grassroots transition process involving lower consumption that emerges differently in different places [3,7,16,45,70]. Examining lived experiences of degrowth from the bottom-up can evaluate whether voluntary simplicity implies a "deep re-evaluation of consumer affluence and embrace of lifestyles of radical material sufficiency" [66] (p. 365). This connects with popular and academic debate in Sweden, signalling a bottom-up resistance against the growth paradigm in multiple crises [29,33,71,72].

Author Contributions: Conceptualization, U.L., M.E. and L.L.; Methodology, U.L., M.E. and L.L.; Software, U.L.; Validation, U.L., L.L. and M.E.; Formal Analysis, U.L.; Investigation, M.E., L.L. and U.L.; Resources, M.E., L.L. and U.L.; Data Curation, U.L.; Writing—Original Draft Preparation, M.E.; Writing—Review & Editing, M.E.; Visualization, M.E. and U.L.; Project Administration, M.E. and L.L.; Funding Acquisition, M.E. and L.L. All authors have read and agreed to the published version of the manuscript.

Funding: This research is supported by the Swedish Research Council for Sustainable Development FORMAS (including funds to cover publication costs, APC) through grant numbers #2016-344 and #2018-547.

Institutional Review Board Statement: Not applicable.

Informed Consent Statement: Not applicable.

Data Availability Statement: Restrictions apply to the availability of these data. Data was obtained from Statistics Sweden through the ASTRID database and are available from the authors with the permission of Statistics Sweden.

Acknowledgments: We are grateful to Charlotta H. for valuable comments and suggestions in many stages of manuscript writing.

Conflicts of Interest: The authors declare no conflict of interest for this article.

References

1. Radetzki, M. *The Green Myth: Economic Growth and the Quality of the Environment*; Multi-Science Publishing Company: Essex, UK, 2001.
2. Thunberg, G. The Disarming Case to Act Right Now on Climate Change. Available online: https://www.ted.com/speakers/greta_thunberg (accessed on 27 April 2021).
3. Nicolosi, E.; Feola, G. Transition in Place: Dynamics, Possibilities, and Constraints. *Geoforum* **2016**, *76*, 153–163. [CrossRef]
4. Lindgren, U. Who is the counter-urban mover? Evidence from the Swedish urban system. *Int. J. Popul. Geogr.* **2003**, *9*, 399–418. [CrossRef]
5. Sandow, E.; Lundholm, E. Which Families move out from Metropolitan Areas? Counterurban Migration and Professions in Sweden. *Eur. Urban Reg. Stud.* **2019**, *27*, 276–289. [CrossRef]
6. Vlasov, M. Ecological Embedding. Stories of Back-to-the-Land Ecopreneurs and Energy Descent. Ph.D. Thesis, Umeå University, Umeå, Sweden, 2020.
7. DeMaria, F.; Kallis, G.; Bakker, K. Geographies of degrowth: Nowtopias, resurgences and the decolonization of imaginaries and places. *Environ. Plan. E Nat. Space* **2019**, *2*, 431–450. [CrossRef]
8. Syssner, J. *Pathways to Demographic Adaptation—Perspectives on Policy and Planning in Depopulating Areas in Northern Europe*; Springer: Cham, Switzerland, 2020.
9. Rebouças, R.; Soares, A.M. Voluntary Simplicity: A Literature Review and Research Agenda. *Int. J. Consum. Stud.* **2021**, *45*, 303–319. [CrossRef]
10. Etzioni, A. Voluntary simplicity: Characterization, select psychological implications, and societal consequences. *J. Econ. Psych.* **1998**, *19*, 619–643.
11. Schor, J. Time, Labour and Consumption: Guest Editor's Introduction. *Time Soc.* **1998**, *7*, 119–127. [CrossRef]
12. Schor, J.B. Sustainable Consumption and Worktime Reduction. *J. Sustain. Ecol.* **2005**, *9*, 37–50. [CrossRef]
13. Tan, P. Leaving the Rat Race to Get a Life: A Study of Midlife Career Downshifting. Ph.D. Thesis, Swinburne University, Melbourne, Australia, 2000.
14. Hamilton, C. *Downshifting in Britain: A Sea-Change in the Pursuit of Happiness*; Discussion Paper 58; The Australia Institute: Canberra, Australia, 2003.
15. Hampton, R.S. Downshifting, Leisure Meanings and Transformation in Leisure. Ph.D. Thesis, Old Main, Pennsylvania State University, State College, PA, USA, 2008.
16. Alexander, S. *Voluntary Simplicity: The Poetic Alternative to Consumer Culture*; Stead and Daughters: Whanganui, New Zealand, 2009.
17. Juniu, S. Downshifting: Regaining the Essence of Leisure. *J. Leis. Res.* **2000**, *32*, 69–73. [CrossRef]
18. Alexander, S.; Ussher, S. The Voluntary Simplicity Movement: A multi-national survey analysis in theoretical context. *J. Consum. Cult.* **2002**, *12*, 66–86. [CrossRef]
19. Beck, U. *Risk Society: Towards a New Modernity*; SAGE: London, UK, 1992.
20. Schor, J.B. *The Overspent American: Upscaling, Downshifting and the New Consumer*; Basic Books: New York, NY, USA, 1998.
21. Lupton, D. Introduction: Risk and Sociocultural Theory. In *Risk and Sociocultural Theory: New Directions and Perspectives*; Lupton, D., Ed.; Cambridge University Press: Cambridge, UK, 1999.
22. Gregory, A.; Milner, S. Editorial: Work-life balance, a Matter of Choice? *Gender Work Organ.* **2009**, *16*, 1–13. [CrossRef]
23. Standing, G. *The Precariat: The New Dangerous Class*, 3rd ed.; Bloomsbury Academic: London, UK; New York, NY, USA, 2016.

24. Hicks, A.; Esping-Andersen, G. The Three Worlds of Welfare Capitalism. *Contemp. Sociol. A J. Rev.* **1991**, *20*, 399–401. [CrossRef]
25. Hedberg, C.; Eimermann, M. Två Forskare om Att Sluta Jobba: Fler Fördelar än Nackdelar Med att Växla Ner. *Aftonbladet*, 18 February 2019; 4–5.
26. Eimermann, M.; Hedberg, C.; Lindgren, U. Downshifting Dutch Rural Tourism Entrepreneurs in Sweden: Challenges, Opportunities and Implications for the Swedish Welfare State. In *Tourism Employment in Nordic Countries: Trends, Practices, and Opportunities*; Walmsley, A., Åberg, K., Blinnikka, P., Jóhannesson, G.T., Eds.; Palgrave Macmillan: Basingstoke, UK, 2020.
27. Eimermann, M.; Hedberg, C.; Nuga, M. Is Downshifting Easier in the Countryside? Focus Group Visions on Individual Sustainability Transitions. In *Dipping in to the North: Living, Working and Traveling in Sparsely Populated Areas*; Lundmark, L., Carson, D.B., Eimermann, M., Eds.; Palgrave Macmillan: Basingstoke, UK, 2020.
28. Eimermann, M.; Lindgren, U.; Lundmark, L.; Zhang, J. Mobility Transitions and Rural Restructuring in Sweden: A Database Study of Holistic Simplifiers. In *Degrowth and Tourism: New Perspectives on Tourism Entrepreneurship, Destinations and Policy*; Hall, C.M., Lundmark, L., Zhang, J., Eds.; Routledge: Abingdon, UK, 2021.
29. Tjärnström, L.A. Ekorrhjulsavhopparna Snyltar på De Som Jobbar och Betalar Skatt. *Göteborgs Posten*. 20 January 2021. Available online: https://www.gp.se/debatt/ekorrhjulsavhopparna-snyltar-p%C3%A5-de-som-jobbar-och-betalar-skatt-1.40121181 (accessed on 6 May 2021).
30. Bastani, A. *Helautomatisk lyxkommunism [Fully Automated Luxury Communism: A Manifesto]*; Verbal Förlag: Stockholm, Sweden, 2019.
31. Hochschild, A. *The Second Shift*; Penguin: New York, NY, USA, 1989.
32. Zimmerman, T.S. Marital Equality and Satisfaction in Stay-At-Home Mother and Stay-At-Home Father Families. *Contemp. Fam. Ther.* **2000**, *22*, 337–354. [CrossRef]
33. Linderborg, Å. Bara Vita Kvinnor får Vara Hemmafruar. *Aftonbladet*. 10 February 2019. Available online: https://www.aftonbladet.se/kultur/a/4dpgde/bara-vita-kvinnor-far-vara-hemmafruar (accessed on 6 May 2021).
34. Boppart, T.; Krusell, P. Labor Supply in the Past, Present, and Future: A Balanced-Growth Perspective. *J. Politics Econ.* **2020**, *128*, 118–157. [CrossRef]
35. United Nations Development Program (2007–2008) Human Development Report. Available online: http://hdr.undp.org/en/content/human-development-report-20078 (accessed on 15 February 2021).
36. Hall, K. Hours Polarisation at the End of the 1990s. *Perspect. Labour Income* **1999**, *2*, 8–37.
37. Ragusa, A.T. Downshifting or Conspicuous Consumption? A Sociological Examination of Treechange as a Manifestation of Slow Culture. In *Culture of the Slow—Social Deceleration in an Accelerated World*; Osbaldiston, N., Ed.; Palgrave Macmillan: Basingstoke, UK, 2013.
38. Schor, J.B. *The Overworked American*; Basic Books: New York, NY, USA, 1999.
39. Morse, E.L.; Inglehart, R. The Silent Revolution: Changing Values and Political Styles among Western Publics. *Foreign Aff.* **1978**, *56*, 442–443. [CrossRef]
40. Brooks, D.; Viladas, P. Inconspicuous Consumption. *New York Times Magazine*. 13 April 1997, p. 25. Available online: https://www.nytimes.com/1997/04/13/magazine/inconspicuous-consumption.html (accessed on 11 February 2021).
41. Breakspear, C.; Hamilton, C. *Getting a Life—Understanding the Downshifting Phenomenon in Australia*; Discussion Paper 62; The Australia Institute: Canberra, Australia, 2004.
42. Peyer, M.; Balderjahn, I.; Seegerbarth, B.; Klemm, A. The Role of Sustainability in Profiling Voluntary Simplifiers. *J. Bus. Res.* **2017**, *70*, 37–43. [CrossRef]
43. Balsa-Budai, N.; Kiss, M.; Kovács, B.; Szakály, Z. Attitudes of Voluntary Simplifier University Students in Hungary. *Sustainability* **2019**, *11*, 1802. [CrossRef]
44. Kennedy, E.H.; Krahn, H.; Krogman, N.T. Downshifting: An Exploration of Motivations, Quality of Life, and Environmental Practices. *Sociol. Forum* **2013**, *28*, 764–783. [CrossRef]
45. Grigsby, M. *Buying Time and Getting By: The Voluntary Simplicity Movement*; State University of New York Press: New York, NY, USA, 2005.
46. Hamilton, C.; Denniss, R. *Affluenza: When too Much Is Never Enough*; Allen & Unwin: Sydney, Australia, 2005.
47. Carbin, M.; Overud, J.; Kvist, E. *Feminism som Lönearbete*; Leopard Förlag: Stockholm, Sweden, 2017.
48. Gunnarsson, A. *Tracing the Women-Friendly Welfare State*; Makadam: Gothenburg, Sweden, 2013.
49. Swedish Social Insurance Agency. Hur Många är Sjukskrivna? Available online: https://www.forsakringskassan.se/statistik/sjuk/sjukpenning-rehabiliteringspenning/hur-manga-ar-sjukskrivna (accessed on 27 April 2021).
50. De Graaf, J.; Wann, D.; Naylor, T.H. *Affluenza: The All-Consuming Epidemic*; Berrett-Koehler: San Francisco, CA, USA, 2001.
51. Stead, D. Urban-Rural Relationships in the West of England. *Built Environ.* **2002**, *28*, 299–310.
52. Massey, D. *Spatial Divisions of Labour: Social Structures and the Geography of Production*; Methuen: New York, NY, USA, 1984.
53. Harvey, D. Roepke Lecture in Economic Geography-Crises, Geographic Disruptions and the Uneven Development of Political Responses. *Econ. Geogr.* **2011**, *87*, 1–22. [CrossRef]
54. Dicken, P. *Global Shift—Mapping the changing Contours of the World Economy*; SAGE: London, UK, 2015.
55. Lindgren, U.; Borggren, J.; Karlsson, S.; Eriksson, R.H.; Timmermans, R.H. Is There an End to the Concentration of Businesses and People? In *Globalisation and Change in Forest Ownership and Forest Use—Natural Resource Management in Transition*; Keskitalo, E.C.H., Ed.; Palgrave MacMillan: Basingstoke, UK, 2017.

56. Gearey, M.; Ravenscroft, N. The Nowtopia of the Riverbank: Elder Environmental Activism. *Environ. Plan. E Nat. Space* **2019**, *2*, 451–464. [CrossRef]
57. Hopkins, R. *The Transition Handbook: From Oil Dependency to Local Resilience*; Green Books: Totnes, UK, 2008.
58. Carlsson, C.; Manning, F. Nowtopia: Strategic Exodus? *Antipode* **2010**, *42*, 924–953. [CrossRef]
59. Massey, D. *For Space*; SAGE: London, UK, 2005.
60. Woods, M. Engaging the global countryside: Globalization, hybridity and the reconstitution of rural place. *Prog. Hum. Geogr.* **2007**, *31*, 485–507. [CrossRef]
61. Hedberg, C.; Do Carmo, R.M. *Translocal Ruralism—Mobility and Connectivity in European Rural Spaces*; Springer: Dordrecht, The Netherlands, 2012.
62. Stockdale, A.; Catney, G. A Life Course Perspective on Urban–Rural Migration: The Importance of the Local Context. *Popul. Space Place* **2014**, *20*, 83–98. [CrossRef]
63. Carson, D.B.; Lundmark, L.; Carson, D.A. The Continuing Advance and Retreat of Rural Settlement in the Northern Inland of Sweden. *J. North Stud.* **2019**, *13*, 7–33.
64. Statistics Sweden. Population Statistics. Available online: https://www.scb.se/en/finding-statistics/statistics-by-subject-area/population/population-composition/population-statistics/ (accessed on 7 June 2021).
65. Hall, C.M.; Müller, D.K. *Tourism, Mobility and Second-Homes: Between Elite Landscape and Common Ground*; Channel View: Buffalo, NY, USA, 2004.
66. Alexander, S.; Gleeson, B. Urban Social Movements and the Degrowth Transition: Towards a Grassroots Theory of Change. *J. Aust. Political Econ.* **2020**, *86*, 355–378.
67. Schierup, C.-U.; Hansen, P.; Castles, S. *Migration, Citizenship, and the European Welfare State*; University Press: Oxford, UK, 2006.
68. Osikominu, J.; Bocken, N. A Voluntary Simplicity Lifestyle: Values, Adoption, Practices and Effects. *Sustainability* **2020**, *12*, 1903. [CrossRef]
69. Löfgren, O. *On Holiday: A History of Vacationing*; University of California Press: Berkeley, CA, USA, 1999.
70. Elgin, D. *Voluntary Simplicity*; William Morrow: New York, NY, USA, 1993.
71. Elander, I. Window of Opportunity or "Here We Go Again"? In *Dipping in to the North: Living, Working and Traveling in Sparsely Populated Areas*; Lundmark, L., Carson, D.B., Eimermann, M., Eds.; Palgrave Macmillan: Basingstoke, UK, 2020.
72. Kallis, G. Societal metabolism, working hours and degrowth: A comment on Sorman and Giampietro. *J. Clean. Prod.* **2013**, *38*, 94–98. [CrossRef]

Hypothesis

An Attack on the Separation of Powers? Strategic Climate Litigation in the Eyes of U.S. Judges

Jasmina Nedevska

Department of Philosophy, Uppsala University, 752 38 Uppsala, Sweden; jasmina.nedevska@statsvet.uu.se

Abstract: Climate change litigation has emerged as a powerful tool as societies steer towards sustainable development. Although the litigation mainly takes place in domestic courts, the implications can be seen as global as specific climate rulings influence courts across national borders. However, while the phenomenon of judicialization is well-known in the social sciences, relatively few have studied issues of legitimacy that arise as climate politics move into courts. A comparatively large part of climate cases have appeared in the United States. This article presents a research plan for a study of judges' opinions and dissents in the United States, regarding the justiciability of strategic climate cases. The purpose is to empirically study how judges navigate a perceived normative conflict—between the litigation and an overarching ideal of separation of powers—in a system marked by checks and balances.

Keywords: climate; litigation; separation of powers; governance; legitimacy

Citation: Nedevska, J. An Attack on the Separation of Powers? Strategic Climate Litigation in the Eyes of U.S. Judges. *Sustainability* **2021**, *13*, 8335. https://doi.org/10.3390/su13158335

Academic Editor: Ingemar Elander

Received: 15 June 2021
Accepted: 20 July 2021
Published: 26 July 2021

Publisher's Note: MDPI stays neutral with regard to jurisdictional claims in published maps and institutional affiliations.

Copyright: © 2021 by the author. Licensee MDPI, Basel, Switzerland. This article is an open access article distributed under the terms and conditions of the Creative Commons Attribution (CC BY) license (https://creativecommons.org/licenses/by/4.0/).

1. Introduction

Pointing to climate-change-related injury, citizens and environmental groups are increasingly turning to courts in order to take legal action against their governments. These cases belong to a category that can be referred to as strategic climate cases. Some of these, such as Massachusetts v EPA (2007) and Juliana v United States (ongoing) in the United States or State of the Netherlands v Urgenda (2015) in the Netherlands, have been the focus of extensive institutional debate [1]. In the Urgenda case, the Supreme Court of the Netherlands established in 2019 that the State's inaction on climate change had violated its citizens' rights to life and privacy [2] (aa. 2; 8). In what has been described as the "strongest" climate ruling so far, the State was ordered to cut its greenhouse gas emissions by at least 25% by 2020, compared to levels in 1990.

Given this development, climate change litigation appears as a potentially powerful tool as societies steer towards sustainable development. Yet, it raises the issue whether, and to what extent, a court may legitimately exercise power over a state's chosen climate policy. To the extent that the aim is to apply specific climate laws, that issue could seem less problematic, albeit not inexistent. Here, however, I aim to treat climate litigation that purports to change government policy, e.g., by reference to fundamental rights–so-called "strategic" climate litigation. Views critical of such litigation are often based on the ideal of separation of powers in constitutional democracies. In a system based on the idea of checks and balances, the separation ideal may allow the judiciary to review legislation, in order to ensure the effectiveness of certain moral constraints. However, the ideal also requires that the judiciary is restrictive in exercising this power.

The United States is a developed system of checks and balances that has seen a comparatively large amount of climate lawsuits. Separation-of-powers principles have also played an important role in U.S. climate change litigation [3] (p. 30). This project uses the United States as a case to investigate a perceived normative conflict between climate change litigation and the ideal of separation of powers. Three studies are suggested in this proposal. Each study comprises a content analysis of opinions and dissents by judges in American climate lawsuits. The particular aims of these are to demonstrate how the conflict (1) is expressed, (2) is avoided by individual judges and (3) is jointly avoided across ideological dividing lines, respectively. While the objects of study are normative views, the particular aims of the studies remain purely descriptive. Nonetheless, the project's findings may be of use for policy-makers and an interested public, including environmental organizations, business representatives, diplomats and lawyers.

2. The Research Problem

From a political science perspective, strategic climate litigation can be seen as a relatively new, overlooked form of climate governance [4,5]. That implies that courts are used to steer entire social systems towards (or away from) sustainable climate policies [6] (p. 385). This would follow a general trend during the late 20th and early 21st century—referred to as the judicialization of politics—to rely on courts to resolve contentious moral matters and public policy issues [7]. Specific climate rulings may, in addition, influence legal considerations in other jurisdictions [8]. Implications of climate change litigation are to this extent "global"—even if it mainly takes place in national courts (for an empirical and normative critique of this view, see [9]).

It seems, furthermore, that the questions raised about the power of courts can be understood through the lens of legitimacy in governance. Authority is said to have normative legitimacy if its claim to power is "well-founded–whether it is justified in some objective sense" [10] (p. 601). Such claims can be made in favour of climate change litigation. Representation of climate plaintiffs in court could be said to contribute to a form of procedural legitimacy: it promotes inclusion in the decision-making process of concerned entities who are excluded from the "demos", i.e., are not eligible voters, such as children, future generations, animals or nature itself [11]. Enforcement by courts, in turn, could be said to contribute to substantial legitimacy: it promotes effectiveness that is lacking due to a relatively unconcerned demos [12,13] and [14] (p. 188). Both goals could be said to promote the effectiveness of climate politics, indirectly or directly. They would thereby be in line with Goal 13 in UN's 2030 Agenda for Sustainable Development: to globally limit climate impact.

However, some will say, the normative legitimacy of climate change litigation does not only rest on its climate-related benefits. Arguably, it should also abide by overarching ideals of constitutional democracies (at least as far as self-proclaimed such states are concerned). To remain with Agenda 2030, it outlines three mutually reinforcing dimensions of sustainable development—an environmental, social and economic dimension, respectively. Goal 16 has a social emphasis and requires responsible, democratic institutions. Less tended to—in the literature on climate change litigation—is the related, institutional ideal of separation of powers (but see [15,16]).

According to this ideal, the power of the different branches of government should be exercised according to the rule of law—a predetermined set of restrictions—in order to reflect the will of the people [17] (pp. 10–13). Judges, defendants and other stakeholders in climate litigation put forth the ideal of separation of powers as a reason to deny strategic climate plaintiffs access to court. For example, the defence in Juliana v United States claimed in the Court of Appeals that such access would be a "direct attack on the separation of powers". (See 18-36082 Kelsey Rose Juliana v. USA-YouTube, at 10:49. Retrieved from the official channel of the United States Court of Appeals for the Ninth Circuit on 27 April 2021.) Part of the expression is included in the title of this project. These critics argue that it is the specified task of elected politicians—not judges far from public control—to steer public climate policy. Here, the legitimacy concern would (again) be procedural, while the specific value is accountability [14] (p. 188). The lack of accountability makes strategic climate litigation seem like a less legitimate form of climate governance.

Nonetheless, very little research has been done on climate litigation in respect of the ideal of separation. It is well known that depending on the constitutional system, the ideal of separation can have radically different implications [17]. In a constitutional system based on the idea of popular sovereignty (such as Sweden, where the present author is from), the separation ideal typically favours the procedural value of accountability through elections, while courts "merely apply the law". Prima facie, that seems to entail that environmentalists should rather seek redress in the parliamentary process. In a system characterized by checks and balances (as in the United States), however, the separation ideal could more plausibly favour the value of effectiveness of some fundamental moral constraints. The judiciary would then be ultimately responsible for keeping the legislator within those constraints, in accordance with a constitution or other basic legal document.

At the same time, judicial constraints on the legislator are to be applied restrictively, in respect of doctrinal criteria. Existing scholarship on rights, in particular, implies that climate plaintiffs may not fulfil such criteria (see, inter alia, [18,19]). Arguably, climate change litigation has left room for individual interpretation among judges. In addition, it is known that judges are influenced by factors external to the law, such as ideological preferences—the law literature refers to the latter as "judicial attitudes". We can expect these to play out more where there is more room for interpretation, as well as in more contentious matters, such as climate policy [20].

It is not clear how judges navigate a percieved normative conflict–between strategic climate litigation and an overarching ideal of separation of powers–in systems of checks and balances. In the project description further below, under the subheading "Material and Methods", I suggest an empirical case study of the United States to investigate this matter.

3. Previous Research

Climate cases have appeared on local, regional, state and federal levels and in a growing number of jurisdictions across the world (including the EU). According to Setzer & Vanhala (2019), research on climate change litigation has proliferated in relation to rulings in high-profile cases such as Massachusetts v EPA (2007) and Urgenda (2015) [5]. In these cases, highest court judges have ruled against defendants (governments) and in favour of strategic plaintiffs on the substantive merits of the case. Yet, very little research has been done on climate change litigation in respect of the overarching ideal of separation of powers. This matter seems understudied both in environmental law and climate governance, which should be the most concerned fields in law and political science, respectively.

Furthermore, few studies on climate change litigation seem to have looked into larger sets of cases, cases on lower levels or dismissed cases (but see [21–25]). This may be, in part, because the high-profile cases contain legal innovations or form precedents for future cases (to some extent, this perspective is kept in the present project). It might also be explained by the fact that climate litigation has been given less attention outside faculties of law.

Under "Material and Methods" below, I suggest a case study of the United States to better understand the complex relationship between strategic climate litigation and the ideal of separation of powers, as far as a system of checks and balances is concerned. The United States offers relatively large amounts of material within a unified legal system, which can be used to describe and explain patterns with regard to the separation ideal. I do so through analyses of judges' arguments [26] in three subordinate studies. In these analyses, I draw on democratic theory, rights theory and scholarship on judicial attitudes, respectively. Rather than provide normative guidelines, the study aims at purely empirical accounts of how a normative ideal (separation of powers) is applied in new legal terrain (strategic climate litigation).

The working hypotheses in the project description partly emerged during a pilot study and postdoctoral appointment at the Department of Politics at Princeton University, 2020. The pilot study involved the applicant and a research assistant under the author's supervision (the author wishes to thank research assistant Akhil Rajasekar).

4. Material and Methods

Against the background of previous research, then, I suggest a specific case study of the United States. The United States is a system of checks and balances with a developed "adversarial legal culture" [11] (p. 12). As such, the United States has seen a comparatively large number of climate lawsuits. As of January 2020, 1143 out of the world's 1444 recorded climate cases had been filed in the U.S. [27]. The cases include, for example, claims based on the National Environmental Policy Act (NEPA), constitutional claims and public trust claims. These cases provide large amounts of material within a unified legal system and can be used to describe and explain patterns of principled argument.

The issue of separation of powers may arise in several ways before a United States court, and this study excludes some. Judges may, for example, refer to separation of powers when a claim is rejected on the merits because there is no relevant right to apply (such as the right to a clean environment) and little room for the judiciary to "invent" new rights [28] (pp. 51, 283–284). This study excludes separation arguments related to merits and focuses on the preceding matter of justiciability (which may include considerations of relevant rights). Justiciability concerns, generally, "a person's ability to claim a remedy before a judicial body when a violation of a right has either occurred or is likely to occur" [3] (p. 30), [29]. A case may be considered injusticiable by reference to a "political question doctrine" if one of the political branches has "a textually demonstrable constitutional" power to resolve the matter or if there are no "judicially discoverable" standards for doing so [30]. The ideal of separation of powers is also reflected in the doctrine of standing. According to this doctrine, plaintiffs need to fulfil established, procedural criteria (judicially discoverable standards) in order to bring their case before the court. Courts tend to invoke the doctrine in order to limit themselves to exercising judicial, rather than legislative or executive, power [3] (p. 30).

The relevant cases can be retrieved online from the Climate Change Litigation Databases at Columbia University's Sabin Center for Climate Change Law. A selection is made of cases brought by strategic climate plaintiffs within the federal legal system. In order to better establish patterns, the project considers the opinions and dissents on several levels for the same case. The project aims to answer the following delimited set of questions:

1. To what extent do U.S. judges express a normative conflict between the litigation and an overarching ideal of separation of powers?
2. How do individual judges navigate the expressed conflict?
3. Do judges' individual views on justiciability of strategic climate cases co-vary with attitudes along ideological lines?

Each question is treated in a study, with the purpose of producing a research article. The overarching method in common for these studies, which are descriptive, is content analysis. The project combines qualitative content analysis (Study 1 and 2) and quantitative content analysis (Study 3) in a manner further described below. Each analysis also draws on a main scholarly discourse—theories on democracy, rights and judicial attitudes, respectively—as described below.

Study 1: The Separation-of-Powers Argument in U.S. Climate Litigation

The first study of the project examines to what extent a theoretical conflict is expressed, in American climate change litigation, between the litigation and an overarching ideal of separation of powers.

Democratic theory is here used to enhance conceptual understanding of the ideal of separation of powers in different constitutional systems (see [17,31,32]) With regard to the U.S. case, a conflict may arise directly, between climate change litigation and the separation ideal, by reference to a political question doctrine, or indirectly, due to a doctrine of standing characteristic of a system of checks and balances. A function of this doctrine is (similarly) to make sure that the judiciary does not exceed its competence in relation to the legislator. In climate litigation, this doctrine may imply several problems, among them a "causality problem", as the plaintiff may not be able to show that injury has potentially been done to them. This study demonstrates the different ways in which strategic climate litigation conflicts with a separation ideal, according to U.S. judges.

Study 2: Three General Approaches U.S. Judges Take to the Causality Problem in Climate Change Litigation

The second study of the project examines how individual U.S. judges navigate an express conflict between strategic climate litigation and the ideal of separation of powers, but only in a further delimited manner. The study I suggest maps the different approaches judges take to one salient problem—the causality problem implied by a doctrine of standing.

Rights theory is here used to distinguish between different approaches to causality (see, inter alia, [5] (p. 10), [24] (p. 40)). A strict individualist approach implies that climate inaction does not cause injury and that plaintiffs are not granted standing, which the first study also shows. Yet, some judges seem to navigate around this by arguing for less established accounts of injury. What I refer to as a lax individualist approach implies that a judge argues for a less established account of individual injury. What I refer to as a non-individualist approach implies that a judge argues that there may be injury of a community (such as a state) or even an object, such as a river [18,19]. I thus hypothesize, with delimited regard to the causality problem, that a conflict persists, between the litigation and a separation of powers, when judges employ a strict individualist interpretation of injury, while judges seek to circumvent the conflict by either a lax individualist or non-individualist interpretation, respectively.

Study 3: Judges' Attitudes in U.S. Climate Change Litigation

The third study of the project explores to what extent judges' views on the justiciability of strategic climate cases co-vary with so-called judicial attitudes (see, inter alia, [25,33–35]).

Research has previously been done on whether judges appointed by Democratic and Republican presidents judge in equal measure in favour of the substantive claims of strategic plaintiffs [25]. Still, no studies of this kind seem to have focused on views on justiciability. However, focusing only on the inclination to hear cases may hide that, depending on attitude, judges' argument(s) for doing (or not doing) so differ. Recent research has also mapped how Republican and Democratic sympathizers in the American populace differ in why they support renewable energy [36]. Here, a similar perspective is used to study judges' views on the justiciability of strategic climate cases.

The issue is explored by using the information in opinions and dissents of individual judges' positions (available through the Sabin Center), together with available data on how each judge was appointed. The complementary data on appointments are available through the Federal Judicial Center's Biographical Directory of Article III Federal Judges [37].

5. Summary: Expected Results

In my first study, I explore the views of U.S. judges on how climate litigation conflicts with the ideal of separation of powers: directly, given a political question doctrine, and indirectly, given a doctrine of legal standing. The latter doctrine seems characteristic for the U.S. as a checks-and-balances system and would make sure that a relatively strong judiciary does not exceed its competence in relation to the legislator.

My second study is limited to a specific but salient problem implied by a doctrine of standing—a causality problem. I hypothesize that three approaches can be identified among judges: a strict, lax and non-individualist interpretation of injury, respectively. Among these, furthermore, I hypothesize that two circumvent the problem and imply that standing is granted: the lax individualist interpretation and the non-individualist interpretation.

In my third study, using additional data on appointments, I explore how views on justiciability co-vary with judges' attitudes along ideological lines. The results in this as well as the preceding studies will, of course, depend on the project's eventual findings.

Funding: This project has received funding from the Swedish Research Council (registration number 2020-06345).

Institutional Review Board Statement: Not applicable.

Informed Consent Statement: Not applicable.

Data Availability Statement: The Climate Change Litigation Databases at the Sabin Center for Climate Change Law at Columbia University: http://climatecasechart.com (accessed on 23 July 2021); Federal Judicial Center's Biographical Directory of Article III Federal Judges: https://www.fjc.gov/history/judges (accessed on 23 July 2021).

Conflicts of Interest: There is no conflict of interest or other ethical objection to this project that the author is aware of.

References

1. Court Cases Retrieved from the Climate Change Litigation Databases at the Sabin Center for Climate Change Law at Columbia University. Available online: http://climatecasechart.com (accessed on 23 July 2021).
2. European Convention of Human Rights (ECHR). Available online: https://www.echr.coe.int/Documents/Convention_Eng.pdf (accessed on 22 September 2020).
3. United Nations Environment Programme (UNEP). *The Status of Climate Change Litigation—A Global Review*; United Nations: Geneva, Switzerland, 2017.
4. Lin, J. Climate Change and the Courts. *Legal Stud.* **2012**, *32*, 35–57. [CrossRef]
5. Setzer, J.; Vanhala, L.C. Climate change litigation: A review of research on courts and litigants in climate governance. *WIREs Clim. Chang.* **2019**, *10*, e580. [CrossRef]
6. Jagers, S.C.; Stripple, J. Climate Governance beyond the State. *Glob. Gov.* **2003**, *9*, 58–78. [CrossRef]
7. Hirschl, R. The Judicialization of Politics. In *The Oxford Handbook of Law and Politics*; Whittington, K.E., Kelemen, D., Caldeira, G.A., Eds.; Oxford University Press: Oxford, UK, 2008.
8. Hirschl, R. *Comparative Matters: The Renaissance of Comparative Constitutional Law*; Oxford University Press: Oxford, UK, 2014.
9. Saunders, C. The Use and Misuse of Comparative Constitutional Law. *Indiana J. Glob. Leg. Stud.* **2006**, *13*, 37–76. [CrossRef]
10. Bodansky, D. The Legitimacy of International Governance: A Coming Challenge for International Environmental Law? *Am. J. Int. Law* **1999**, *93*, 596–624. [CrossRef]
11. Epstein, Y. The Big Bad EU? Species Protection and European Federalism. A Case Study of Wolf Conservation and Contestation in Sweden. Ph.D. Thesis, Uppsala University, Uppsala, Sweden, May 2017.
12. Hellner, A. Arguments for Access to Justice. Supra-Individual Environmental Claims before Administrative Courts. Ph.D. Thesis, Uppsala University, Uppsala, Sweden, October 2019.
13. Scharpf, F.W. *Governing in Europe: Effective and Democratic?* Oxford University Press: Oxford, UK, 1999.
14. Nasiritousi, N.; Verhaegen, S. Disentangling Legitimacy. Comparing Stakeholder Assessments of Five Key Climate and Energy Governance Institutions. In *Governing the Climate-Energy Nexus*; Zelli, F., Backstrand, K., Nasiritousi, N., Skovgaard, J., Widerberg, O., Eds.; Cambridge University Press: Cambridge, UK, 2020.
15. Burgers, L. Should Judges Make Climate Change Law? *Transnatl. Environ. Law* **2020**, *9*, 55–75. [CrossRef]
16. Bergkamp, L.; Hanekamp, J.C. Climate change litigation against states: The perils of court-made climate policies. *Eur. Energy Environ. Law Rev.* **2015**, *24*, 102.
17. Hermansson, J. Om att Tämja Folkmakten. Maktdelning, Demokratiutredningens Forskarvolym I (SOU 1999:76), ed. Amnå. 1999. Available online: https://data.riksdagen.se/fil/8E1E1145-2B37-4BFA-9A87-1390FBC8861D (accessed on 23 July 2021).
18. Taylor, C. *Philosophy and the Human Sciences: Philosophical Papers 2*; Cambridge University Press: Cambridge, UK, 1985.
19. Stone, C.D. *Should Trees Have Standing? Law, Morality and the Environment*; Oxford University Press: Oxford, UK, 2010.
20. Segal, J.; Spaeth, H. *The Supreme Court and the Attitudinal Model. Revisited*; Cambridge University Press: Cambridge, UK, 2002.
21. Markell, D.; Ruhl, J.B. An empirical assessment of climate change in the courts: A new jurisprudence or business as usual? *Fla. Law Rev.* **2012**, *64*, 15. [CrossRef]
22. Bogojevic, S. EU Climate Change Litigation. *Law Policy* **2013**, *35*, 184–207. [CrossRef]
23. Osofsky, H.M.; Peel, J. The Role of Litigation in Multilevel Climate Change Governance: Possibilities for a Lower Carbon Future? *Environ. Plan. Law J.* **2014**, *30*, 303.
24. Peel, J.; Osofsky, H.M. Climate change litigation's regulatory pathways: A comparative analysis of the United States and Australia. *Law Policy* **2013**, *35*, 150–183. [CrossRef]
25. Keele, D.M. Climate change litigation and the national environmental policy act. *J. Environ. Law* **2018**, *30*, 285–309. [CrossRef]
26. Parsons, C. *How to Map Arguments in Political Science*; Oxford University Press: Oxford, UK, 2007.
27. Norton, R.F.; de Wit, E.; Seneviratne, S.; Calford, H. Climate Change Litigation Update. Available online: https://www.nortonrosefulbright.com/en-it/knowledge/publications/7d58ae66/climate-change-litigation-update (accessed on 6 April 2020).
28. Boyd, D.R. *The Environmental Rights Revolution: A Global Study of Constitutions, Human Rights, and the Environment*; UBC Press: Vancouver, BC, Canada, 2011.
29. International Commission of Jurists. Courts and the Legal Enforcement of Economic, Social and Cultural Rights (ICJ): Comparative Experiences of Justiciability. Available online: https://perma.cc/YU9F-YCNR,2008 (accessed on 23 July 2021).
30. Anonymous. Harvard Law Review. Political Questions, Public Rights, and Sovereign Immunity. *Harv. Law Rev.* **2016**, *723*, 130.
31. Bellamy, R. The Political Form of the Constitution: The Separation of Powers, Rights and Representative Democracy. *Polit. Stud.* **1996**, *44*, 436–456. [CrossRef]

32. Nergelius, J. Grundlagsmodeller och maktdelning. Maktdelning, Demokratiutredningens forskarvolym I (SOU 1999:76), ed. Amnå. 1999. Available online: https://data.riksdagen.se/fil/8E1E1145-2B37-4BFA-9A87-1390FBC8861D (accessed on 23 July 2021).
33. Sunstein, C.; Schkade, D.; Ellman, L.M.; Sawicki, A. *Are Judges Political? An. Empirical Analysis of the Federal Judiciary*; Brookings Institution Press: Washington, DC, USA, 2006.
34. Anzie Nelson, L. Delineating Deference to Agency Science: Doctrine or Political Ideology? *Environ. Law* **2010**, *40*, 1057.
35. Jylhä, K.; Akrami, N. Ideology and climate change denial. *Personal. Individ. Differ.* **2014**, *70*, 62–65.
36. Gustafson, A.; Goldberg, M.H.; Kotcher, J.E.; Rosenthal, S.A.; Maibach, E.W.; Ballew, M.T.; Leiserowitz, A. Republicans and Democrats differ in why they support renewable energy. *Energy Policy* **2020**, *141*, 111448. [CrossRef]
37. Biographical Directory of Article III Federal Judges. Federal Judicial Center. Available online: https://www.fjc.gov/history/judges (accessed on 23 July 2021).

MDPI
St. Alban-Anlage 66
4052 Basel
Switzerland
Tel. +41 61 683 77 34
Fax +41 61 302 89 18
www.mdpi.com

Sustainability Editorial Office
E-mail: sustainability@mdpi.com
www.mdpi.com/journal/sustainability

www.ingramcontent.com/pod-product-compliance
Lightning Source LLC
LaVergne TN
LVHW070636100526
838202LV00012B/819